FROM PARADOX TO REALITY

From paradox to reality

OUR BASIC CONCEPTS OF THE PHYSICAL WORLD

FRITZ ROHRLICH

Syracuse University

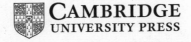

CAMBRIDGE
UNIVERSITY PRESS

PUBLISHED BY THE PRESS SYNDICATE OF THE UNIVERSITY OF CAMBRIDGE
The Pitt Building, Trumpington Street, Cambridge CB2 1RP, United Kingdom

CAMBRIDGE UNIVERSITY PRESS
The Edinburgh Building, Cambridge CB2 2RU, United Kingdom
40 West 20th Street, New York, NY 10011–4211, USA
10 Stamford Road, Oakleigh, Melbourne 3166, Australia

First published 1987
First paperback edition 1989
Reprinted 1990, 1992, 1997

Printed in the United Kingdom at the University Press, Cambridge

A catalogue record for this book is available from the British Library

Library of Congress Cataloguing in Publication data

Rohrlich, Fritz
From paradox to reality.
Bibliography
Includes indexes.
1. Relativity (Physics)   2. Quantum theory.
3. Science – Philosophy.   I. Title.
QC173.55.R64   1987   530.1′1   86–26845

ISBN 0 521 30749 X hardback
ISBN 0 521 37605 X paperback

Contents

Preface

Come, join me on an adventure of the mind. The going will be a little demanding at times but there will be rich rewards. The new vistas are spectacular. Come with me, I am an experienced guide!

I am offering the reader a guided tour to the two new fundamental conceptual structures of twentieth-century physical science: relativity and quantum mechanics. This is *not* a physics text; nor is it an account of the latest discoveries intended to dazzle the public. It is a basis for discussion of surprising new concepts and of the necessities that led to them.

This book developed from my lecture notes for a one-semester course that I have been offering to undergraduates at Syracuse University for a number of years. The course is entitled 'Concepts in Contemporary Physics'. While I have taught almost every branch of theoretical physics, I find this course to be by far the most demanding. One reason is that the course is intended for non-science majors; a serious interest in the subject matter is the only prerequisite. Consequently, the students have widely varying backgrounds and range from freshmen to seniors. Another reason is the lack of a suitable text. A reading list pulling together bits and pieces from a variety of sources does not replace an integrated text. The present book is expected to fill this need.

However, the book is not restricted to use by college students. It is intended to attract a much wider audience: the intellectually interested and more sophisticated of the general public. And I hope it will serve another goal: to correct widespread misrepresentations of science. While there exist many excellent popular presentations of various fields of science, the public has in recent years also been exposed to a large number of books on various pseudo-sciences. Since these range widely from astrology to creationism, there also resulted a great deal of confusion and misunderstanding concerning the basic physical sciences. The present book, though limited in its scope, is expected to help dispel misunderstandings in at least two frequently misunderstood topics: relativity and quantum mechanics.

The book consists of three parts: Part A brings together some basic notions from the philosophy of science that provide the background and prepare the reader for the two main parts; Part B deals with the revolution in our concepts of space and time brought about by the two theories of relativity, the theory of very fast motion (special relativity) and the theory of gravitation (general relativity); and Part C presents the greatest conceptual revolution of all, the discovery of the quantum world and the resulting new world view.

Most of the sections can be read without recalling one's high school mathematics. In a few of them, however, a little elementary algebra or a few basic geometric notions will be helpful. But those sections can be skipped with little ill effect for the understanding of later material. The two parts on the world of relativity (Part B) and on the quantum world (Part C) are largely independent and either of them can be read first.

Human desires for solutions to problems and for certainty in a world of uncertainty often make people reach out for easy answers. But science does not have easy answers and often has no answers yet. It requires patience and sometimes a long wait for further scientific progress. But where scientific answers are still lacking, the door is open to wild speculations; people then often fall prey to freak ideas. One of these is the introduction of Eastern philosophy into the foundations of quantum mechanics. Both are respected accomplishments of human thought; but they have nothing to do with one another. The present book is in part intended to provide sufficient understanding of this field of science to avoid such fallacies.

The conceptual structure of science raises not only intellectual curiosity but also philosophical issues that are often of much greater interest to the non-scientist than the subject matter itself. Correspondingly, this book is not a popularization of the subject matter of twentieth-century fundamental physics. It does not attempt to teach relativity or quantum mechanics or even to present its subject matter in popularized form. There already exist a large number of books for this purpose, many of them authoritative and excellent. My present endeavor is rather to present in non-technical language the *conceptual revolutions* that the scientific community had to undergo in order to be able to accept 'modern physics'. And this is done less from a historical point of view than from the perspective of a person confronted with new, strange, and often paradoxical experimental evidence. The general public as well as the college student hears little of that unless he or she pursues a scientific career. But it is exactly this type of intellectual adventure that shows scientific research at its best. It teaches what science is all about and it permits one to develop a world view based on scientific evidence.

If there is just one thing the reader should learn from the following pages, I hope it will be this. Our common sense, acquired from our everyday experience, is not a reliable guide to the world far beyond our reach in daily life. Nature is found to be much more subtle, varied, and sophisticated than we tend to imagine. Our prejudices about how nature *should* behave are usually wrong. It is this experience that gives scientists their humility; more than the non-scientist, they are often painfully aware of how much we do not know and how difficult it is to acquire reliable knowledge. This is why they treasure so highly the relatively small amount of reliable knowledge which we did manage to acquire.

The nature and purpose of this work require me to deal not with science only; some philosophy of science will play a necessary role. For example, we shall encounter philosophical issues in connection with the interpretation of quantum mechanics. Two philosophical points of view, instrumentalism and realism, will provide the minimum background for this discussion. My own view, that of a critical scientific realist, will unavoidably enter the picture. But I suspect this view to be close to what the majority of scientists today actually believe.

Side remarks that would interrupt the flow of thought are relegated to Notes at the end of the text; they are indicated by square brackets. The technical terms are collected in a Glossary for easy reference. Finally, lists of further readings are provided at the end of Part A and after each of the subsequent chapters. They are in the form of alphabetical annotated bibliographies and are generally limited to non-mathematical material. Of course, they reflect my own taste and preference. Instructors may want to add their own favorite references.

As is the case in any introductory text, some simplifications were necessary. While every effort has been made to avoid oversimplifications in either science, philosophy, or history, the expert will find small liberties that I had to take in order to avoid devoting excessive space to very minor points. In some instances the Notes provide elaborations for the sake of precision. Similarly, matters that are fascinating but had to be excluded because of limitations in length are indicated in the Notes. There is of course a large amount of personal judgment involved in the choice of subject matter. Items that others may have liked to include had to be omitted entirely in order to keep the book to its present size. It is already considerably larger than had been intended originally.

I owe a great debt to the many authors whose works I have read over the years and who have influenced me greatly. They are too many to list here. But I must express my special indebtedness to Abner Shimony, Professor of Philosophy and Physics, Boston University, for the many things I have learned from him. My great indebtedness goes to two people who have provided valuable constructive criticism: Clyde L. Hardin, Professor of

Philosophy, Syracuse University, who has patiently read through earlier versions of nearly the entire book, and R. Bruce Martin, Professor of Chemistry, University of Virginia, who read earlier versions of most of Part C. I want to thank them here in a public way. Any deficiencies in the final version of this book are of course entirely my own doing. I have also profited greatly from my students who, over the years, have asked many pertinent and baffling questions. Sometimes, they made me rethink what I thought I had understood. And last but not least, my thanks go to our librarian, Mrs Eileen Snyder, whose patient and unfailing help has been invaluable.

Syracuse, New York   F. R.
June 1986

Part A

At the root of the endeavor

The most incomprehensible fact about nature is that it is comprehensible.
Albert Einstein

It is not *what* the man of science believes that distinguishes him, but *how* and *why* he believes it. His beliefs are tentative, not dogmatic; they are based on evidence, not on authority or intuition.
Bertrand Russell

1

Human limitations

There are various limitations and restrictions with which scientific investigations must cope. Three of these will be mentioned here very briefly. They will recur at different places in our story. All of them are deeply interwoven in our recent quest for scientific knowledge.

a. *Limits by principle*

We human beings are part of nature. This trivial observation has profound consequences for both philosophy and science: the pursuit of scientific investigation can in a sense be looked at as nature trying to study itself. One can surmise in this view that difficulties may lie ahead.

Before the present century it was taken for granted that the study of nature is in no way affected by the process of measuring. If one is only clever enough one can measure things to arbitrary accuracy. There may of course be better and worse investigators; different people may use more or less sophisticated methods. But the results should, in principle, not be influenced by such things. Psychologists who study human behavior are of course aware of the fact that people behave differently when they know that they are being observed. But no such thing can happen with inanimate matter.

The scientific revolutions of our present century taught us otherwise. Quantum mechanics, the theory involving the laws of nature on the atomic and subatomic level, tells us that every observation influences the things we observe; that there is a limit to the accuracy that can be achieved in a measurement; and that these matters have nothing to do with the ingenuity of the observer or with the technological sophistication of his or her apparatus. These are limitations by principle.

That is shattering news. The great scientific and technical achievements of the eighteenth and nineteenth centuries pointed to unbounded progress in exploring nature to satisfy our curiosity. We were thereby also assured of learning to control our environment for our benefit. And now we are told that there are principles in nature that tell us: so far and no further!

There is a fundamental question that arises here: do these newly discovered principles only limit our *knowledge* of nature, making for a 'blurred' knowledge, while true reality is actually sharp but not accessible to us? Or, does true reality itself actually exist in a blurred or fuzzy state? The large majority of physicists now believe the latter: in the atomic world properties of objects can be blurred in a way completely unknown to us from our everyday experience. No wonder that this question has been occupying scientists and philosophers of science ever since the development of the mechanics of the atom (quantum mechanics). We shall deal with it at length in Chapters 10 and 11.

But let us not misunderstand the new insight. The lessons of quantum mechanics do not put limits on the possibility of progress; they only direct progress in telling us what is not possible. For instance, the location of an atom at a given instant of time and its speed at that same time cannot simultaneously have exact (sharp) values. This is a matter of *principle* (Section 10d); it lies in the nature of things. We are just not used to such limitations because they do not exist in our macroscopic world (Section 10f).

Vision is perhaps the most developed of our senses. The object we see may emit light of its own, like a firefly at night, or it may be illuminated by some light source such as a lamp or the sun. In either case it is necessary that light leaves the object and reaches our eye where it is focused by the eye's lens. Eventually, it arrives at the light sensitive layer, the retina, which activates the brain. Thus we cannot sense an object unless it sends us a message; in the case of vision, that message is in the form of light. The object must therefore 'do something' in order for us to know about it. It must either produce and send light to us, or it must reflect light from some source: the object must be actively involved. And it must do so sufficiently strongly because our eyes are not sensitive below a certain light intensity.

Consider the firefly. It must expend a certain amount of energy, no matter how small, in order to produce light and send it to us. That energy is contained in the light that reaches our eyes; it activates the retina. The interaction between the firefly and us requires that energy. By the time we see that firefly it has already expended some of its energy, no matter how little, in order to send us its light. It had more energy before we saw it than afterwards.

While this is perhaps a trivial example, it makes the point: our observation requires some change to take place in the object we observe. By the time we observe the object it is different (perhaps ever so slightly) from what it was before we observed it. *The observation affects and changes the observed.*

Of course, in everyday life these changes of observed objects are usually much, much too small to be noticeable. But on the atomic level they play a crucial role (Chapters 10 and 11).

☐ Firefly is different, but would be w/o
observation, too. It still emits light.
Objects do change during observation,
but because of us?

b. *Bounds of human nature*

Our human bodies can survive and function only within very narrow bounds. Tolerable temperatures range only over about 50 degrees centigrade, and even that is tolerable only with suitable clothing. But the temperatures in nature range from very near absolute zero (−273 degrees centigrade) to many millions of degrees in the center of stars. We require air pressure in our surroundings of just the right amount within very narrow limits; some people begin to feel dizzy at the top of a mountain. But the pressures that occur in nature range from near zero to over one million times atmospheric pressure, even within our own earth, not to speak of the interior of stars. Our eyes are sensitive only to a very narrow band of wave lengths (or frequencies) of light, the visible range. We cannot see X-rays, or radio waves, or in fact most radiation. We need a very special kind of gas to breath, air of just the right mixture of oxygen and nitrogen. And one can go on and on citing examples.

Granting these narrow bounds outside of which human existence is not possible, how then do we acquire knowledge about that world that lies outside these narrow bounds? Contrary to the limits by principle which we discussed above, these can be overcome. What we need, however, are suitable devices that in some sense 'translate' or convert information from the world outside our livable range into information that we can perceive directly. Thus we are dependent on perception aids; we are forced into *indirect perception*.

Temperature is a good example. By means of a thermometer we can determine much higher temperatures (boiling water) than those in which we could survive. A thermometer can convert temperature into the length of a column of mercury. And with that we have no troubles at all because it is within our range and can be perceived directly.

A much less trivial example is the science of astronomy. For all the many centuries up to the present all our knowledge about the stars came from visual observations. Whether people used the naked eye or sophisticated telescopes, all information was received in the form of visible light. Only relatively recently did we discover that there are 'messages' reaching us from the sky in the form of radio waves (1933) and in the form of X-rays (1948). We were simply not aware of those because we could not see them. But when people devised suitable detection devices which convert radio waves and X-rays into visible messages new 'windows in the sky' were opened to us. Whole new branches of astronomy were launched: radio-astronomy and X-ray-astronomy. We learned to 'see' previously invisible radiation, and thus were enabled to study the many sources in space that emit such signals.

In a similar way we have extended our natural vision below the smallest objects we can see with our naked eyes: we have developed the optical microscope. And more recently we have topped that by the invention (in 1931) of the electron microscope, and (just a few years ago) of the scanning tunneling microscope. The latter permits 'seeing' of individual atoms [1.1].

All these are ways of exceeding the bounds imposed on us by human nature. They involve indirect perception, and they demonstrate how the indirectness increases as we venture out further into the unknown.

The presence of narrow bounds outside of which humans are unable to function but nevertheless wish to explore nature raises another important issue. Our knowledge of the world as we experience it in everyday life establishes our *common sense*. We have a certain feeling about what is reasonable and what is not reasonable. And here we run into our natural tendency to extrapolate, to assume that the world outside our usual bounds is similar to the world we are used to. We tend to be *prejudiced* about what to expect; we believe that we know what nature is going to be like even in new and heretofore unknown domains. And we are all too often wrong in that.

One of our main goals will be to show how different matters can be outside our common experience, and how our prejudice has indeed led us astray repeatedly. We must be prepared to enter worlds where common sense is a poor guide. The first example we shall encounter is the world of very high speeds, speeds close to the speed of light. That is the world of special relativity, as developed by Einstein and others after him (Chapter 6). Another one will be the world of gravitation where we shall find that our notions of space and time no longer apply when astronomical sizes are involved (Chapter 7). And finally we shall deal with the most difficult world, the world of quantum mechanics. There, the conversion from the atomic and subatomic worlds to the world to which we are accustomed produces a clash of seemingly contradictory notions. We shall indeed be faced with apparent paradoxes. The story of how scientists struggled to overcome them is a fascinating tale. It will occupy us in Chapter 11, and it will lead us to a new notion of reality, different, strange, and – beautiful.

c. *Restrictions by complexity*

The last type of limitations to human capabilities is related to the complexity of nature. People have learned to cope with complexity by idealization and abstraction. They would ignore certain features of the problematic phenomena under study which are considered irrelevant details. They would thereby idealize them so that their models of nature would no longer agree with reality in all respects. The results are *approximations* to reality. Usually these are excellent approximations as the following examples will demonstrate. And the simplification achieved thereby permits the desired solution to complex problems. Such idealizations are ubiquitous in scientific as well as in non-scientific activities:

A traffic engineer is studying the flow of large numbers of cars that hurry through the main arteries of a big city during rush-hour. He approximates each car by a small rectangle on the big city map. The make of the car, its color, the type of engine, etc., are all entirely irrelevant to him. He has replaced the actual system by an idealized one which in no way will invalidate the conclusions of his study. On the contrary, it will help his work when he eliminates unimportant details.

An astronomer studies the dynamics of the solar system. His records of observation of the motion of the planets show them as points in space. Their finite diameters, the substances they consist of, etc., are all ignored. But he does not ignore their masses or the forces acting among the planets and between the planets and the sun. His 'model' of the solar system thus involves such absurd and unreal notions as points that have a mass. Nevertheless, this model gives exactly the correct description of planetary motion. It even permits correct predictions such as the exact time of the next solar eclipse.

The best way to characterize these situations is to introduce the notion of *levels of reality*. What is ignored on the level of the traffic engineer is essential on the level of the auto mechanic; what is ignored on the level of the astronomer is essential on the level of the geologist. Nor will these different levels lead to contradictions. Our earth is a mass point for the astronomer but surely not for us. We and the astronomer simply operate on different levels.

The world of science can be neatly broken down into these levels of reality. By going increasingly deeper into the subject matter we have, in order of increasing attention to detail, the following examples of different fields of study: anatomy, molecular biology, atomic physics, and nuclear physics. The subject matter of each of the last three is essentially ignored or idealized almost beyond recognition by the field preceding it on the list. We shall return to such levels of reality in Chapter 8 and especially in Section 12b.

Which of these levels, however, is the *true* nature of reality? Is the true understanding of nature to be found on the deepest level [1.2]? The answer to this often debated question is that *all* levels of reality are equally true. They just describe different approximations. Each level presents features of reality not found on other levels. For example, atomic physics tells us nothing about the nature of life which we learn from the study of living organisms.

But there is also the progress of science which seems to continue to find deeper levels. We have no assurance that there is a deepest level. Thus, only when all levels are considered together, i.e. when we have before us nature in all approximations, do we obtain a complete picture or at least a picture as complete as present-day science would admit.

A little contemplation now makes us aware of the fact that we are actually *always* dealing with approximations. *Science is approximate on every level.* When we talk about the 'exact' sciences we surely cannot mean exactness in the sense of lack of idealization. Approximation by idealization is an essential part of 'sorting things out', of simplifying the complexity of nature so that we can handle it. Whether this is good or bad is not the question. It is necessary.

This awareness of approximations goes hand in hand with our restricted ability to measure. Every measurement has some error. And it is therefore meaningless to attempt greater accuracy than the given level of reality would admit. We all know that our earth is more or less spherical. It is certainly not flat as was believed by most people at least as recently as the time of Columbus. Yet, when we draw the lines for the foundations of a house we pay no attention to the curvature of the earth. We do it as if the earth were flat. And there is of course no problem with that: the accuracy required in constructing a house is much too poor to detect the curvature of the earth.

The same is true for scientific theories. They are also associated with approximations that are 'good enough' for the level of reality at hand. We admire the perfectly smooth surfaces of a highly polished perfectly ground photographic lens, although we know very well that on a deeper level that surface consists of atoms and molecules and that it could not possibly be as smooth as we claim it to be. But for the problem at hand (perhaps the sharp focusing of sunlight reflected from the petals of a yellow rose onto a photographic film) this is an excellent approximation. And that is all we can ask.

We conclude with the remark that it is not always so clear which features of a particular phenomenon are essential and which features can be safely ignored. The great scientists that extracted laws of nature from the tremendous complexity of the world around us clearly succeeded in distinguishing the important from the irrelevant. The rare ability to do that requires uncommon insight and may well be more like an art than a science.

2

Theory and the role of mathematics

a. *What is a scientific theory?*

There is a large variety of meanings in the common usage of the word 'theory'. One often hears 'that's true in theory but it's not true actually', or 'that's true in theory but not in practice'. When Jack wonders why the paper boy missed his delivery on certain days he may say, 'I have a theory about that.' These are not the meanings of 'theory' that we want to use.

We are concerned here with scientific theories and in particular with those of the physical sciences. Those theories are of a very special kind; they are the result of long and careful study of a particular class of phenomena in nature carried out by many scientists. They involve *models of reality* that have been confirmed by numerous observations [2.1]. They are no longer conjectures or hypotheses but have passed that preliminary stage, although they may still not yet be well established. In the physical sciences such theories are usually associated with mathematical descriptions that may in some instances be quite extensive and abstract. In the latter case one speaks of the *mathematical structure* of a theory.

As an example of a physical theory consider *geometrical optics*. It is the theory of that branch of physics that deals with light as it penetrates various 'media' such as air, water, glass, etc. In geometrical optics light is pictured as a ray, a pencil of light, that is produced when a small opening is put in front of a light source. That ray traces a curve as it moves through transparent substances. It is the 'model of reality', the idealization, of the actual thing in this particular theory. When one holds a magnifying glass between the sun and a piece of paper held at the correct distance, the sun's rays will focus on the paper and burn it. One can think of each ray as a single line. The straight rays of the sun are then seen to be 'broken' by the lens in such a way that they all intersect exactly at the focal point of the lens. Thereby they combine all their energy at that point and allow the burning to take place.

The theory permits one to compute these lines or 'paths' of the light rays by a principle known as *Fermat's Principle*. It is the fundamental principle on which that theory is based. It asserts that the actual path of a light ray between two points takes less time than any neighboring path would take that starts and ends at the same two points [2.2]. The shape of that path can be determined mathematically. In this way one can find out, for example, how light rays move through the lenses of optical instruments. One can also determine the bending of light rays in the earth's atmosphere due to the change of air density with altitude. This bending of light rays causes a flattening of the sun's circular disk at sunrise and sunset.

Similarly, the laws of reflection (Fig. 2.1) and refraction (Fig. 2.2) of light [2.3] emerge as consequences of Fermat's Principle. This provides an example of how various specific laws can be deduced from a scientific theory. Of course, the theory was constructed so that this is the case because these laws were known earlier. They provided valuable input for the construction of the theory. The theory now combines them and unifies them into one logical mathematical structure so that they can be derived from that one principle.

The value of a theory lies of course not only in its ability to reproduce the laws out of which it was constructed but also in its ability to predict new laws or to explain newly discovered phenomena. We shall deal with that in Section 4a.

A very appropriate question to raise is how one knows whether a theory is correct. The history of science shows that many theories which were accepted at one time had to be abandoned later because of experimental evidence against them. Of course, this was the case especially in the eighteenth and nineteenth centuries when experiments were not so reliable and were often inaccurate. At that time theories were accepted as correct on what we today would consider to be rather insufficient evidence. A famous example is the

Fig. 2.1. Reflection of a light ray.

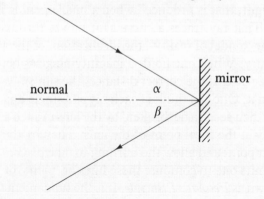

caloric theory of heat that was generally accepted in the eighteenth century. It pictured heat as a fluid called 'caloric'. It survived for only about a generation but lost credibility when accurate experiments (mostly by Benjamin Thompson – Count Rumford (1753–1814)) proved that this caloric (a) has no weight, and (b) can be generated seemingly inexhaustibly by friction. It could therefore not possibly be a fluid.

Better experiments of course do not assure us that one or other of our current theories may not also have to be abandoned at some later time. To guard against that, various criteria can be drawn up that should be met by a satisfactory theory. Such acceptability criteria will be considered in Section 3b.

But there were also instances where theories were *superseded* by seemingly better theories. Those new theories could involve completely different models of reality. The old theories would therefore be considered wrong and obsolete, and many people thought that they should be abandoned. These *scientific revolutions* will be considered in detail in Chapter 8; but we want to emphasize already here that the old theories are in general not being abandoned, nor should they be. If an old theory has been well established and has served us well and successfully then it is likely that it is a *good approximation* to the new theory. In that case, it is usually much more convenient to use *it* instead of the new one in many instances. We shall deal with this matter much more in Chapter 8.

b. *What does mathematics have to do with nature?*

It may seem rather surprising that rigorous mathematics would have anything at all to do with nature. The richness and variety of natural phenomena is overwhelming. It would seem absurd that behind it all there are mathematically precise laws. Yet this is exactly what physical scientists claim.

Fig. 2.2. Refraction of a light ray.

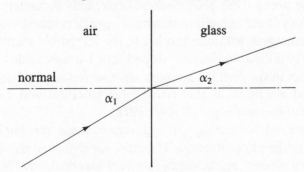

How do they know? They don't know, at first. They start by assuming that rigorous mathematics is applicable and after long and hard work they do find that their assumption was justified. This means that physical laws are found to be correct not only qualitatively but even quantitatively. In fact, mathematics was found to play an increasingly important role not only in physics but also in related fields such as astronomy.

Once the *mathematical foundations* of a physical theory have been formulated one is forced to accept its mathematical consequences lest the whole structure becomes logically inconsistent. For example, the mathematical formulation of Fermat's Principle leads by means of mathematical steps to the predictions for the paths of light rays in different optical instruments such as microscopes and telescopes. If these predictions were not borne out after construction of these instruments the whole theory based on that principle would have to be abandoned.

But the mathematical consequences of a theory's fundamental principles lead not only to predictions for specific situations; they also lead to *scientific laws* which are derived mathematically from those principles. The laws of optical reflection and refraction are examples [2.3]. Applying these laws permits one to *predict* how an optical instrument is going to function. This, in turn, permits one to design them according to specifications. One sees how *applied science* depends heavily on the laws discovered by *fundamental science*.

The mathematical derivation of laws from fundamental principles uses algebra, calculus, and other higher branches of mathematics. But when it comes down to a numerical comparison between the theory and experiments arithmetic is called for. Another aspect of the importance of mathematics then becomes evident: the incredible numerical accuracy to which certain theories have been confirmed. Everyone knows of the ability of astronomers to predict a solar eclipse many years in advance of its occurrence, and the precision of this prediction in date and exact time is indeed most remarkable. Less well known but equally impressive are predictions of electromagnetic properties of certain elementary particles. The electron behaves like a little magnet whose strength has been measured extremely accurately. The corresponding theory (quantum electrodynamics) predicts this strength. Theory and experiment agree with one another to the incredible accuracy of more than 1 part in 10 billion. In a lecture entitled 'The Unreasonable Effectiveness of Mathematics in the Natural Sciences' the theoretical physicist Eugene P. Wigner called attention to this remarkable situation and called it 'the empirical law of epistemology' (Wigner 1978).

There is indeed something quite uncanny about the applicability of mathematics to the physical world. The more sophisticated physical theories of this century contain mathematics in a very essential way. Mathematics, then, almost becomes part of the model. In a way, it takes over and guides us toward the right way of looking at reality. This is of the greatest importance for the epistemology and ontology of scientific theories.

The mathematical formulation of a theory introduces mathematics as a kind of 'language'. Concepts are characterized by mathematical symbols and are manipulated as such. But of much greater interest is the converse: mathematical symbols are given an interpretation in terms of concepts. These may or may not be new concepts, they may or may not be related to our everyday experience. The result is that *the mathematical formulation of a theory contributes to the interpretation of the model of reality*. The results can be quite unexpected and sometimes even rather strange, as in the case of quantum mechanics, the theory of the atomic and subatomic world. The abstract nature of mathematics can add a new dimension to our thinking that allows us to go far beyond our everyday experience. This fascinating feature is in fact one of the main themes of our journey into the realm of the scientific quest. We shall encounter it in the theory of relativity and even more so in quantum mechanics.

This power of mathematics is closely related to the use of rather sophisticated branches of it in physical theories. Of course, we shall not use advanced mathematics here; but we shall point to specific instances where mathematics is able to connect apparently completely different and unrelated notions, and where it leads us far beyond our human bounds and prejudices that we discussed in Chapter 1. And if one contemplates the problem of exceeding our human bounds we can find little else that can guide us in the right direction. A purely verbal account of reality would suffer badly under the ambiguity of everyday language, the variation in denotative as well as the connotative meanings of words causing considerable difficulties. This is evident in the sciences that have so far had only little possibility to use mathematics to great advantage. No wonder that those sciences that did have that possibility have been able to make much faster progress.

The question of the flexibility of a theory is often raised: can a theory be modified to adjust for a discrepancy with experiments? In a theory that is still 'in the making' this is of course done all the time. But once a mathematical structure has been adopted this flexibility is very greatly reduced. We see here another role of mathematics in a scientific theory: it has a strongly stabilizing effect; it prevents the theory from being 'faked' into agreement with empirical results.

Finally, there is also great beauty in a physical theory. Unfortunately such beauty is not easily accessible to those who are not sufficiently well acquainted with the subject. The deeper one enters into it the more one learns to appreciate it. The situation is similar to that of music: the more one hears it the more one learns to love it. The beauty of a scientific theory is to a considerable extent due to its mathematical structure. Again, such mathematical beauty is often reserved for the experts in the field. But it is that beauty which affects the credibility of one theory over another in the absence of more stringent criteria. For instance, the general theory of relativity is so beautiful that it is preferred over rival theories as long as those rival theories cannot account any better for the empirical facts.

Mathematics thus plays a variety of roles in scientific theories. In general terms it provides precision and rigor for a theory. But more specifically it endows it with the ability to deduce many different laws from only a very small number of fundamental principles; it permits the computation of specific predictions; it allows great numerical accuracy; and it prevents ambiguities and faking of agreement with data. It also endows a theory with a certain beauty and 'elegance' which is unfortunately not always apparent to the non-expert. But most importantly, in mathematically advanced theories, it provides a guide to the interpretation of the theories that helps us to exceed our human bounds for the exploration of the worlds beyond. It leads us to models of reality far beyond our common experience.

3

Scientific objectivity

a. *The evolution of a scientific theory*

Scientific inquiry, like any other inquiry, stems in the last analysis from curiosity. Therefore, like anyone else who is curious without being able to obtain an answer right away, the scientist speculates. He asks nature particular questions which may or may not be the correct or the important ones. If his best hypothesis fails he will be forced to try another one. This is why two scientists may attack the same problem in entirely different ways. Subjective preferences, personalities, individual abilities and technical skills play a vital role.

Science 'evolves' rather than 'is developed' in the sense that its progress resembles much more a Darwinian evolution than a systematic or pro-grammed development. A scientific theory in the making is full of trials and errors, of hunches and of hypotheses that are too often enthusiastically embraced only to be abandoned again later, often reluctantly. It is a battle-ground of the imagination, of one scientist's wit against another's. It's the struggle for survival of the fittest theory.

A lot of the effort of finding an appropriate theory, including a suitable mathematical formulation, is thus a matter of educated guess, of intuition, and of imagination. Despite considerable talk about it (which has largely historical reasons) there is no such thing as *the* scientific method which one can simply follow in order to make one's discoveries. Methods of inquiry differ widely from one discipline to the next. They depend as much on the available technology as on the ingenuity of the investigator. Two people working on the same problem may use entirely different methods in order to achieve the same goal. If one looks for prescriptions that are *generally* valid one finds only such obvious demands as reliability of tests and procedures, critical attitude, rational analysis, etc. Most of the activity, however, is simply hard work. The old adage is still true: it takes 5% of inspiration and 95% of perspiration to succeed [3.1].

Upon inspection of the actual evolution of a scientific theory one finds that theorizing and experimentation are going on simultaneously, checking and supporting one another. There is a steady give and take between them. It is impossible to design an experiment beyond the most elementary level without the use of a certain amount of theory in the the preparation of it; and it is impossible to construct a substantive theory without having at least some empirical facts available or, preferably, one or more laws that have a good chance of being correct. The quest for a scientific understanding is laden with covert and overt assumptions, working hypotheses, which guide the individual researcher along his or her particular way. No general systematic procedure exists that one can safely follow.

How, then, does one ensure scientific objectivity? First of all there is a feedback mechanism that keeps the individual 'honest'. I do not mean that this mechanism is needed to prevent a scientist from wanton misrepresentation, though such instances have also been known to occur. Rather, it is needed to force the individual to guard against unintentional mistakes. A complex experiment may contain unsuspected systematic errors; a long computation may hide numerical errors despite repeated checks. For these reasons the results of an individual or a single research group cannot be generally accepted without reservations. Those same results must be obtained independently by a different researcher or research group at a different place using different methods and different instrumentation. In this way science is a *self-correcting* activity.

Then there is the fact that the scientist is extremely cautious in accepting a new result as well as in discarding an old belief. He is basically very conservative in this respect. Accepting new ideas can have an adverse effect on him since he does not want to be proven wrong. On the other hand, he wants to be the first one to find something that others later on (after checking it by means of their own work) will have to accept. Consequently, the scientist is always looking for new ideas and results but seldom willing to take them seriously unless he has very strong reasons for believing them.

One of the obstacles to objectivity in the evolution of scientific theories has played an especially important role in the past. It is the acceptance of such theories on the basis of *authority*. There exist well-known examples of that in the history of science. The authority of Aristotle and Ptolemy dominated the scientific view of the Church until the time of Galileo and beyond. It resulted in a very slow acceptance of the Copernican revolution that called for the replacement of the geocentric system by his heliocentric one. As is well known, Galileo who accepted the Copernican system was called to Rome and had to recant before the Inquisition (1633). This last period of Galileo's life has become notorious in the history of intellectual freedom.

The authority of Newton overshadowed Huygens' wave theory in favor of Newton's corpuscular theory of light (1704), even though Newton himself stated that *both* theories are able to account for the observed phenomena. The wave theory won out eventually [3.2].

A more recent example involves the authority of a government. In the Soviet Union the work by the agriculturist Trofim Lysenko was supported by the authority of the state. It became the officially accepted scientific view (Lysenkoism). Lysenko's belief in the inheritance of acquired characteristics contradicted the opinions of most of the experts in genetics and heredity both inside and outside of the Soviet Union. Lysenkoism had disastrous effects on individual scientists in the Soviet Union as well as on Soviet agriculture. It had to be abandoned eventually [3.3].

Efforts to ensure objectivity go a long way to establish sound foundations for a scientific theory. But these alone cannot guarantee that the theory itself is correct. In addition, criteria of acceptability are necessary. We must review them briefly. Apart from their intrinsic value they indicate the tremendous amount of meticulous work that must be expended before a theory can receive general acceptance.

b. *Acceptability criteria*

There is no acceptability criterion or combination of criteria that is foolproof. No theory can ever be *proven* to be correct in the same sense as a mathematical theorem can be proven. The best one can expect is to be convinced of its correctness with overwhelmingly high probability [3.4]. To this end various criteria are available. A list of such criteria which most scientists will agree upon as at least important, if not necessary for acceptability, might look like the following.

(1) *Empirical confirmation*: a scientific theory must permit confirmation or disconfirmation by experiment or by observation. Some theories may have only very few but quite accurate data of confirmation. This was the case (until recently) with Einstein's gravitation theory (the general theory of relativity, Chapter 7) when there were only three confirming observations. In other cases a large amount of data may be available but these may be of much poorer quality.

However, one must be careful: an experiment or an observation that contradicts a theory does not necessarily mean the immediate demise of that theory. (The skepticism that scientists bring to new and seemingly revolutionary results has already been mentioned (Section 3a).) It lies, at least partly, in the existence of several alternatives: the empirical data may be incorrect due to *errors in the measurements*; or the theory may be correct but the *computations* that were made to deduce this specific consequence from the theory *were in error*; or, finally, the data and the theory may be correct, but the theory is being probed outside its *validity limits*.

In the latter case the theory may *seem* to fail because those validity limits are not yet known. An example is the failure of theories of pre-quantum physics (called classical physics) at the turn of this century. They simply were unable to explain a variety of well-established phenomena (Section 9a). This failure of the classical theories led eventually to the discovery of quantum mechanics. That new theory in turn provided certain validity limits for the classical theories. It thereby explained why these classical theories were unable to account for those phenomena; it exempted them from that duty and thus permitted them to continue as viable theories as long as they are restricted to their own validity domain. The validity domain of classical theories had not been known previously. (The question of validity domains will occupy us at length in Chapter 8.)

In order to permit empirical confirmation a theory must be able to *predict*. A scientific theory that does not permit predictions is consequently highly questionable. The more unexpected the prediction (if confirmed) the better for the theory. Einstein's gravitation theory predicted the bending of light rays by gravitation. Its confirmation (not only qualitatively but also quantitatively) during the solar eclipse of 1919 removed a great deal of the still existing doubt and made that theory acceptable to the satisfaction of the large majority of the scientific community.

(2) *Consistency with other theories*: many phenomena are in the sphere of competence of more than one field of science. For example, the age of pottery shards and of other archaeological findings can be determined in several different ways: one can ascertain the age of the geological layer in which they are found; one can match them with similar pieces of a culture of known age; or one can use radioactive dating methods. These different dating methods must all be consistent with one another. Now radioactive dating is based on the theories of nuclear physics so that if these results were to be considerably different from those obtained by the other methods nuclear physics would become questionable. That type of consistency is referred to as *horizontal consistency*.

There is another type of consistency, *vertical consistency*. It arises when theory A is more general than theory B. This means that theory A has a domain of validity that contains within it the domain of validity of theory B. Consistency in that case requires that theory B can be *deduced* from theory A when the domain of theory A is restricted to that of theory B. One speaks of *reduction* of theory B to theory A. One example is the electromagnetic theory of light (as theory A) from which one deduces geometrical optics (theory B) in the limit of sufficiently small wave lengths [3.5]. Other examples will be encountered in Chapters 6, 7, and 10. We shall discuss reductionism in Section 8b.

(3) *Simplicity and beauty*: these may be surprising as criteria for a scientific theory because they are necessarily based on subjective perception. But we have already encountered beauty as an asset in mathematics (Section 2b). A considerable amount of scientific theory may also be judged beautiful. The problem lies in the difficulty that a non-scientist has in appreciating this quality of a scientific theory.

Simplicity, like beauty, can usually also be appreciated only by the expert. But both are used consciously or unconsciously by physical scientists. The theoretical physicist Paul A. M. Dirac, for example, claimed to have been guided in his research to a considerable extent by considerations of beauty and simplicity. To what extent this guidance dominated is of course very difficult to tell. But he certainly has been tremendously successful; his achievements won him the Nobel prize in 1933 at the relatively young age of 31.

A theory may be accepted by the scientific community if it satisfies acceptability criteria such as the above to a sufficient degree. They help greatly to guard against belief in an incorrect theory. History has shown repeatedly that theories whose correctness was believed widely had to be discarded. One such example is the caloric theory of heat (Section 2a). Another is phlogiston theory [3.6]. Although most of these examples date from scientific generations of the not so recent past, there is no guarantee that one of our presently accepted theories may not also share the same fate. It is therefore useful to distinguish theories that are merely *accepted* from theories that can be considered as *established*. The latter are theories that are extremely likely to be correct and are believed to be very close to certainty (see [3.4]). We shall devote more time to established theories after we have encountered some (Chapter 8); the theories whose concepts we shall study at some length, special and general relativity theory and quantum mechanics, are such established theories.

c. *Invariance: the irrelevance of the specific observer*

So far 'objectivity' has been considered in various meanings of the word. But there is yet another sense in which objectivity can be meant, especially in the context of the physical sciences. It derives from the question of generality of the laws of nature. Are they valid for everyone? This seems to be a superfluous question since one would not want to dignify a statement with the words 'law of nature' if it were not true for everyone who cares to take the time and trouble to make the appropriate observations or measurements.

Yet, matters are not that simple. One way in which two experimenters may differ is in their *state of motion*. Will two observers who differ in the way in which they move relative to the phenomena they investigate find the same laws of nature?

By 'observer' is meant here the person who is making the observation or measurements. But what matters here is clearly not that person but his motion, the motion of his 'laboratory' relative to what he observes. That laboratory or, more generally, that 'location of the observer' to which all observations are referred is called a *reference frame*. Such a reference frame can be thought of as any idealized rigid structure that is at rest relative to the observer. It is in general not part of the object which is being studied. If a rocket that has been launched to the moon is observed from an observation post on the ground below, or from a moving car, or from a moving plane, we speak of three different reference frames. Observations can be made relative to each one of them. Are the laws of nature the same no matter which reference frame is used to do the observations?

If one charts the motion of an object such as that rocket one can draw the shape of its trajectory, the line along which it moves. One can also mark on that trajectory at what locations the rocket was at different times. If this is done relative to one reference frame (by an observer on the ground, say) and also relative to another one (by an observer orbiting the earth in a spaceship, say) will they find the same results? Will they find the same shape of the trajectory for that rocket? Surely, the two observers will find *different* shape trajectories. Does this mean that different laws of nature are applicable for observations made relative to different reference frames? The answer must surely be 'no' but it seems far from obvious how to relate different observations like these to the same natural laws. That is the problem of objectivity we wish to raise now.

The solution to this problem has occupied physicists extensively since the time of Newton. Three different solutions have been found, each of greater generality than the preceding one. All of them specify the conditions under which observers in different reference frames can use the same laws of nature. The first solution, *Newtonian physics*, is valid only for speeds that are small compared to the speed of light. The second one, *Einstein's special theory of relativity*, removes this restriction and is valid for all attainable speeds. But both of these solutions require that no forces be acting on the reference frames. The third one, *Einstein's gravitation theory* (the general theory of relativity), removes that restriction in that it allows at least gravitational forces to act on the reference frames. In that theory the same laws of nature hold for all reference frames no matter what their speed and irrespective of their being subject to gravitational forces. These matters will occupy us at length in Part B.

Within these restrictions the laws of nature are the same relative to all reference frames. Equivalently, one can say that the laws of nature are independent of the reference frame to which they are referred. This property is called *invariance*. The reason for this term is the following. As one looks from the data taken relative to one reference frame to those taken relative to another (both having recorded the same phenomena) one must 'convert' or 'transform' one set of data into the other set. For example, the trajectory of that rocket must be transformed from the shape seen relative to the ground to the shape seen relative to the orbiting spaceship. This requires taking account of the difference in the motion of the two reference frames. That transformation will not, however, affect the laws of nature since the same laws are found in both. The laws of nature are therefore unchanged or *invariant* under these transformations.

Since these problems are intimately related to space and time the invariance with respect to reference frames is often called *space–time invariance*.

There are other invariance properties of the laws of nature which deserve mention at this point. But these do not tell us about the generality of the laws of nature with respect to motion. They do not refer to transformations between two frames but, rather, to transformations of a very different sort. When we look into a mirror and compare the mirror image with ourselves we can also relate the two by a transformation. This transformation changes left into right and right into left. Do the laws of nature change under such a transformation? We know this *not* to be the case because anything that takes place as seen in the mirror can also take place in reality. And that was scientific 'dogma' until quite recently. Yet, in 1957 it was discovered that while this is indeed true for the electromagnetic forces and those acting in atomic nuclei, the weak forces of nature (see Section 4b) *do* change when seen in a mirror: not *everything* that occurs as seen in a mirror can also occur in reality [3.7]. We have thus learned to specify which laws remain invariant under certain transformations (like the mirror transformation) and which ones do not.

There is a close relationship between invariance and *symmetry*. Consider a geometrical figure. A square has a certain symmetry. One consequence of this symmetry is the invariance of that square under rotations. Such rotations must be carried out about its center and in the plane of the square. After a rotation by 90 degrees the square will be positioned as before and therefore look exactly as it looked before the rotation. Such rotations therefore leave the square invariant. If we were to do the same thing with a regular hexagon instead of a square we would rotate by 60 degrees instead of 90. In this way the relation between symmetry and invariance can be exploited to characterize the particular kind of symmetry an object has.

For this purpose one must make a list of all the transformations that would leave the object invariant. Different symmetries are associated with different such lists. One then studies how the different transformations on the list are related to one another. From such studies emerges a characterization of different kinds of symmetry. The science of crystallography uses this method to classify the different shapes of crystals. The particular branch of mathematics that provides that link between symmetry and invariance is called *group theory* because that list of transformations is called a transformation group. It permits a sophisticated way of studying the properties of the transformations on that list. Group theory plays a leading role in twentieth-century physical sciences [3.8].

Invariance of the laws of nature and symmetry in nature are thus very closely related. They are part of the ingredients that matter to scientists who judge a theory to be beautiful. The fact that nature exhibits beauty as a result of symmetry is of course well known to anyone who has ever admired a flower. But one does not always remember that this symmetry is a result of the laws that were active when that flower developed. Indeed, symmetry lies very deep in nature (Hermann Weyl, 1952).

4

The aim of scientific theory

a. *Explanation and prediction*

Anyone who has tried to explain something is aware of the need to express himself so that he relates the unknown which is to be explained to something that is known by the questioner. Otherwise, true understanding is not possible. This is, however, not always possible in scientific explanations. About a century and a half ago, when a farmer saw his first railroad train pass by his fields he wanted to know how that train could move without being pulled by horses. But he could not obtain an explanation in terms of something he already knew. The answer 'a steam engine makes the wheels turn' was meaningless to him and did not provide him with true understanding since he did not know what a steam engine was. He had first to be taught about steam engines and only then could he be given the above explanation. The unknown could not be reduced to something known to him since the steam locomotive was too recent an invention. Similarly, recent scientific theories have first to be made known to the non-scientist before they can be used for the purpose of explanation.

The usual scientific explanation reduces something specific to something more general. The specific instance may be the question why a mirror has to be tilted in a certain way in order to see something that is taking place behind you. The general law governing this instance is the law of reflection of light (Section 2a).

But a law can also be explained. It can be reduced to a more general construct, a theory. The law of reflection is explained by geometrical optics (Section 2a). Geometrical optics can in turn be reduced to a still more general theory; it can be explained in terms of the electromagnetic theory of light. Thus, there are explanations of instances by laws, explanations of laws by theories and, in certain cases, even explanations of theories by more general theories.

Historically speaking, in all these cases the more general item had to be discovered before the explanation became possible. The farmer had to be told about the steam engine before he could understand what makes the train move. The progress of science involves a steady increase in our knowledge of laws and from these we build theories so that we can explain laws in terms of them; and we use theories to build more general theories in the hope that we can deduce more and more knowledge from fewer and fewer very general theories.

In this process the important matter of approximation must not be overlooked. We encountered the matter of approximation earlier (Section 1c). It permits one to explain one level of description of reality by another one which is more general and therefore considered to be more basic. Since that more basic description is usually also more sophisticated it is often referred to as a 'higher level description' or a 'higher level theory'. Here are some examples: geometrical optics is an approximation to electromagnetic theory; Newton's theory of gravitation is an approximation to Einstein's gravitation theory (the general theory of relativity, Chapter 7); classical physics is an approximation to quantum physics (Section 10f). In all these cases the higher level theory 'explains' the lower level one. But careful inspection reveals that such explanations are not of the same nature as those that explain instances by laws or laws by theories: different levels of reality involve different conceptual structures (different logical or scientific categories) and approximations become a necessity when one deduces a lower level theory from a higher level one (Chapter 8).

Most scientists believe that they explore nature as an external reality. It exists and functions quite independent of our observations and in fact independent of our very existence. Scientific theories are *descriptions of that reality*: while there is only one reality, it has many descriptions ('faces'). This view is called the *realist view*. But there are other possible points of view in the philosophy of science. One other view is of special interest to us. This view is called the *instrumentalist view*.

According to the instrumentalist view a scientific theory plays the role of a tool, of an instrument, of a computational 'device'. The only things that are accepted as certainly real are the results of measurements and observations. A theory relates these realities, connecting them by some logical or mathematical structure. Theory is thus used only as a means to an end. The concepts involved in it may in general have no reality, no significance as representations of real objects, they may not refer to anything at all. They are intermediary concepts that help establish a connection, an interpolation between things that are observed. The theory is just a useful construct, and no realistic interpretation of its content is assured. Correspondingly, an explanation by means of a theory is according to the instrumentalist view *not a description of reality*.

If a fundamental electrically charged particle such as an electron crosses a photographic film it leaves behind a track. This track appears in the developed film as a line which is microscopically a series of grains of silver. That is what one really sees. The electron itself is not seen. Its existence is inferred. The theory says that as the electron passes through the photographic emulsion it forces other electrons out of atoms in the emulsion. These in turn trigger processes that eventually lead to the deposit of pure silver grains after development. The instrumentalist view gives credence only to the observed track but is at best non-committal on the actual existence of that electron that started it all. The realist view does not question the reality of that electron; it was really there, it existed.

The branch of philosophy concerned with questions of being is called *ontology*. The instrumentalist and the realist view of science differ in the ontology of nature. The instrumentalist is non-committal while the realist insists on the existence of the objects which the theory describes.

Physical scientists have on the whole paid little attention to instrumentalism until the mid-twenties when quantum mechanics was developed. Under the leadership of the Danish physicist Niels Bohr and the German physicist Werner Heisenberg, a certain amount of instrumentalism was injected into the accepted view for the interpretation of quantum mechanics. It became known as the *Copenhagen interpretation* because Niels Bohr who was the senior scientist among those who were primarily responsible for this development was active at the Institute of Theoretical Physics in Copenhagen, Denmark. His contributions had a dominant influence on the interpretation problem of quantum mechanics. Werner Heisenberg was one of the younger physicists who visited that Institute for extended periods. He was one of the founders of quantum mechanics but he also contributed substantially to the interpretation problem.

The instrumentalist interpretation was developed at a time when philosophy of science in Europe was under the strong influence of a view known as *logical positivism*. This view was very sympathetic to instrumentalism since it championed empiricism [4.1]. Many of its main proponents lived at that time in Vienna, Austria, where it flourished in the 1920s and 1930s (they became known as the 'Vienna Circle'). After the Second World War, criticism of logical positivism increased at the same time as its philosophical program was unable to produce the anticipated results. As a consequence, the attractiveness of logical positivism declined over the years so that by the early 1970s it was no longer a viable doctrine in the philosophy of science. This trend was also reflected by increasing questioning of the philosophical soundness of the Copenhagen interpretation. The issues raised became more and more a matter of an instrumentalist versus a realist interpretation of quantum mechanics. While many issues have been resolved in the intervening years the matter is still under investigation. Sections 11d and 11e are devoted to this problem.

One of the vital features of a scientific theory is its ability to predict what will happen under suitably well-specified circumstances. Using the mathematical apparatus of a theory to compute the outcome of an experiment that has not yet been performed seems to be very similar to the computation of the outcome of an experiment that *has* been carried out and whose results are already known. In the first case one deals with a *prediction*. In the second case, when the theory is in agreement with the results, one can regard the computation as an *explanation* of the experimental result in terms of that theory. Thus, there seems to be a symmetry between explanation and prediction.

However, this symmetry may not always be present because of the peculiar nature of a particular theory. The theory of evolution permits excellent explanations of the variety of species of plants and animals, but it is difficult to make predictions. An example of the opposite extreme is the astronomical theory of Ptolemy (2nd century AD). He placed the earth in the center of the universe and described the motion of the sun and the planets in a sufficiently complicated way so that he was able to *predict* their motion to very high accuracy. For instance, he could predict solar and lunar eclipses. But his system was so complex that it could hardly be considered to provide an *explanation* of the motion of the sun and the planets. It is simply not 'understandable' in the way in which one wants an explanation to be. (As a parenthetical remark, Ptolemy himself never meant it to be explanatory; he was in a sense an early instrumentalist.)

The predictive capacity of a theory can provide more than numerical predictions. It can lead to new concepts and to the prediction of entirely unexpected phenomena. A well-known example is the laser which is basically an instrument which produces very intense and very well-collimated radiation. It is based on a phenomenon discovered by Einstein in 1917 called 'stimulated emission' [4.2]. He predicted this phenomenon on the basis of the early quantum theory. If one accepts the instrumentalist view of a theory which regards it as a computational tool only, then it becomes difficult to see how such a new phenomenon can be predicted. One suspects that a theory is a great deal more than an instrument for relating observations. On this basis one would therefore favor the realist view over the instrumentalist one.

The interplay between explanations and predictions is a recurrent theme throughout the physical sciences. Its scientific importance increases with the number of kinds of phenomena explained or predicted; and these can be qualitative or quantitative explanations and predictions. Thus one finds that one often explains and predicts whole *classes* of phenomena. Newton's gravitation theory accounts not only for the motion of the earth around the sun but also for the motion of all the other planets, as well as for the motion of the moon around the earth, the moons of Jupiter around Jupiter, etc. And the larger and more varied the class of phenomena explained or predicted by a theory, the more powerful that theory is considered to be. The aim of the physical sciences is therefore in this direction also: one wants to construct more and more powerful theories accounting for more and more phenomena.

b. *Unification*

In a very crude way one can describe the progress of physical science as proceeding from experiments and observations to the discovery of laws, and from a collection of laws to the construction of theories. These are processes of unification. Beyond that one is sometimes able to find a theory from which previously known theories can be deduced as special cases or as approximations. One then also speaks of unification.

The process of unification permits one to understand a larger domain of phenomena than has been possible before by means of fewer fundamental principles or theories. One accomplishes thereby a deeper understanding which is of course a goal of science in general. The present section is intended primarily to provide examples of unification. It will also permit us to gain an overview of the fundamental physical forces and of the theories which deal with them.

The discovery of quantum phenomena and the subsequent development of quantum mechanics led to the realization that the quantum world is not only quantitatively but even *qualitatively* entirely different from anything that physical science has encountered before. Consequently, one distinguishes the quantum world from the other one by calling the latter 'the classical world'. One speaks of *quantum physics* and *classical physics*, of quantum theories and of classical theories which describe, respectively, quantum phenomena and classical phenomena. The three hundred years of physical science since Galileo Galilei (1564–1642) was all classical physics, and so are the theories of special and general relativity developed early in the present century. The understanding of atoms and molecules (which includes almost all of modern chemistry) is based on quantum theories. And so is our knowledge of the atomic nucleus, nuclear science, and subnuclear science, and the study of fundamental particles. But there are also macroscopic phenomena that depend crucially on quantum theories for their understanding. These include lasers, superconductivity, and the world of semiconductors which underlies the possibility of modern computers. We shall elaborate on some of these in Section 11g.

One of the earliest unifications was the discovery by Johannes Kepler that all six planets known at the time obey the *same* three laws concerning their motion. This was a unification of a very large body of observational data taken over a long period of time and by a variety of astronomers, especially Tycho Brahe. These laws proved correct also for the planets that were discovered later on. At about the same time the laws of motion for bodies falling under the earth's force of gravity were studied by Galileo. Both Kepler's and Galileo's laws constituted a unification of observations into laws.

The first great unification of laws into a theory was the achievement of Isaac Newton. His theory of universal gravitation (1687) combined the laws of Kepler with Galileo's laws of motion under the earth's gravitational attraction. But it went far beyond these laws because it provided small corrections to these laws thus specifying the approximation to which they are correct.

One cannot underestimate the tremendous revelation it must have been for people to learn that the same law of gravitation (of Newton's theory) underlies *both* the heavenly motions of planets and the earthly motions of falling apples (Fig. 4.1).

Another equally impressive unification took place about two hundred years later (1873) when James Clerk Maxwell unified the large body of accumulated knowledge of electricity, magnetism, and light. Many laws had been discovered over the years dealing with electric and magnetic phenomena, and the science of optics had been extended ever since the fundamental discoveries by Newton of the properties of light (his famous book '*Opticks*' first appeared in 1704). Maxwell's electromagnetic theory showed the intimate interconnection between electric and magnetic phenomena, and established (together with suitable experiments mainly by Heinrich Hertz) the electromagnetic nature of light (Fig. 4.1).

Fig. 4.1. Unification of classical phenomena into two fundamental theories.

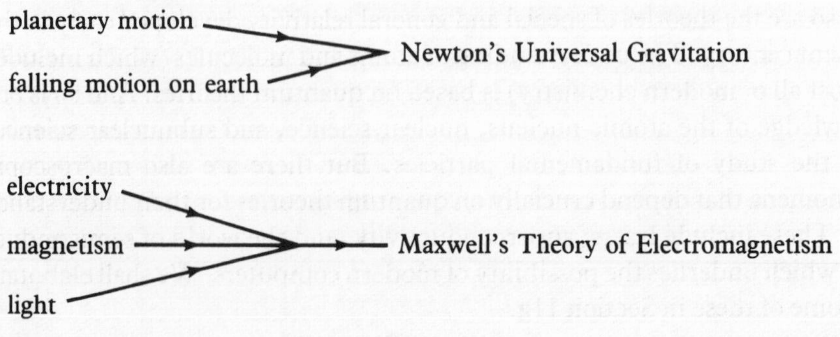

planetary motion ⟶
                    ⟶ Newton's Universal Gravitation
falling motion on earth ⟶

electricity ⟶
magnetism ⟶   ⟶ Maxwell's Theory of Electromagnetism
light ⟶

We shall return to this matter in greater detail in Section 6a. But the unification that was brought about by the discovery of the electromagnetic nature of light as well as of many other types of radiation is most remarkable and must be told at this point. These types of radiation, some of which were not discovered until long after Maxwell, differ greatly in their appearance and effects: heat radiation, radio waves, and X-rays are all electromagnetic waves and differ only in their wave length (see Fig. 4.2). Visible light occupies only a tiny range of wave lengths of the enormous electromagnetic spectrum. This tiny range is the only one to which our eyes happen to be sensitive.

The unifications brought about by Newton and Maxwell completed the discoveries of the fundamental forces of the classical world. We know today that all of the many and varied phenomena of classical physical sciences can be reduced on the fundamental level to just *two* basic forces: the *force of gravitation* and the *electromagnetic force*. This, too, is a most remarkable insight afforded by the process of scientific unification.

One may ask why we do not have a further unification which would combine the theories of Newton and Maxwell. We must defer the answer to this question to a later time (Section 7d).

Let us now turn to the quantum world. Here the discoveries of recent times established two new fundamental forces, forces that do not exist in the classical world. For lack of a better name these have become known as the *weak force* and the *strong force*.

The weak force was first met when radioactivity was discovered. It is responsible for the decay of certain atomic nuclei. These nuclei break up into somewhat smaller atomic nuclei and other fundamental particles such as electrons, alpha-particles, or neutrinos. Alpha-particles are the nuclei of helium atoms; neutrinos are very strange elementary particles that have neither an electric charge nor a mass and that always move with the speed of light.

Fig. 4.2. The electromagnetic spectrum. The notation 10^n is explained in Note 4.3.

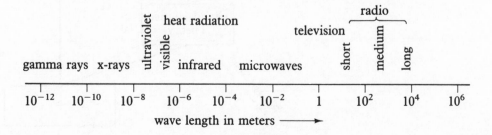

The strong force is responsible for holding the atomic nuclei together. As its name implies, it is indeed very strong. As a consequence, the energy contained in an atomic nucleus due to the strong binding force is enormous. It is this energy that provides the radiation energy of the sun and the stars – as well as the explosive power of nuclear weapons.

In addition to the two characteristic quantum mechanical forces, the weak and the strong forces, there are only two other fundamental forces to the best of our present knowledge. These are the quantum versions of the two classical forces, the quantum-electromagnetic, and the quantum-gravitational forces. The theory of the quantum-gravitational force is still in a very conjectural stage of development.

The last example of unification concerns these four fundamental forces (Fig. 4.3). So far only the unification of the quantum-electrodynamic and the weak forces has succeeded. The resultant theory is known as the theory of the *electroweak* force. This theory became established only quite recently [4.4].

Much of the forefront of current research in the theory of fundamental particles is therefore now devoted to the unification of the strong force with the electroweak one. The last force, the quantum-gravitational force, is probably not going to be unified with the others for some time to come.

Fig. 4.3. The unification of the four fundamental interactions. The dashed line indicates the approximate present state of the art.

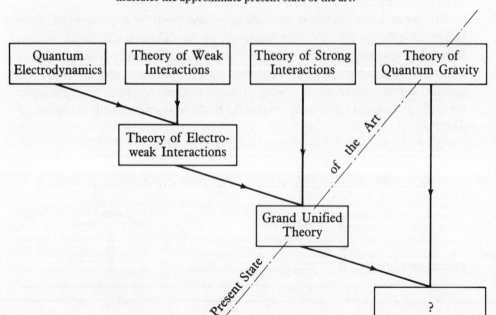

However, in the last several years a very bold conjecture has attracted a lot of attention. It is the proposal of a theory which would combine not only electroweak and strong forces but would also include quantum gravity. This conjecture is called 'string theory'. As of this writing it has no empirical support whatever, but exists only as a mathematical construct.

Figures 4.1 and 4.3 provide a brief overview of the relations between our theories of fundamental forces. String theory would fit into the empty box of Figure 4.3.

Annotated reading list for Part A

Asimov, I. 1972. *The Left Hand of the Electron*. New York: Doubleday Co. The first part of this book by a master of popularization of science contains an account of the left–right symmetry in nature and of the discovery of its violation.

Duhem, P. 1962. *The Aim and Structure of Physical Theory*. New York: Atheneum. A classic on the subject written around the turn of the century by a physicist with a strong interest in the history and philosophy of science. His view is that of an instrumentalist.

Gardner, M. 1957. *Fads and Fallacies in the Name of Science*. New York: Dover. The author is well known as a popularizer of mathematics and the physical sciences. His article on Lysenkoism in this little volume is of interest in connection with the question of authority in science.

Poincaré, H. 1952. *Science and Hypothesis*. New York: Dover. This is one of several collections of popular essays by one of the leading mathematicians of the late nineteenth century. See especially Chapter IX.

Weyl, H. 1952. *Symmetry*. Princeton: University Press. A beautiful and well-illustrated book written in non-technical language. Weyl was one of the great mathematicians who also contributed to the application of mathematics to the sciences.

Wigner, E. P. 1978. 'The unreasonable effectiveness of mathematics in the natural sciences'. In *Symmetries and Reflections, Scientific Essays of E.P. Wigner*. Bloomington: Indiana University Press. This Nobel laureate in physics is best known for his work on invariance and symmetry in physics using the mathematical methods of group theory.

Ziman, J. 1978. *Reliable Knowledge, An Exploration of the Grounds for Belief in Science*. Cambridge: University Press. A highly respected theoretical physicist who also champions the case for the appreciation of the human, cultural, and social influences on the scientific enterprise. Some scientists feel he is overstating the case.

Part B

The world of relativity

The book of Nature is written in the mathematical language.
Galileo Galilei

Henceforth space by itself and time by itself are doomed to fade away into mere shadows, and only a kind of union of the two will preserve an independent reality.
Hermann Minkowski

5

Space and time: from absolute to relative

This chapter serves as background to the world of relativity. Both the special theory of relativity and the general theory of relativity are creations of our present century. One cannot understand these theories fully without some notion of what had gone on before. At the same time it is impossible within the present work to do justice to the long and fascinating intellectual history that led from the time of Galileo via Newton and his successors to the emancipation of the study of motion, resulting in the science of mechanics as it stood at the beginning of the present century. Its history is in large part the story of physical science in general as it came into its own during that heroic time known as the period of 'classical science'.

We shall start with a brief review of those great contributions by Galileo that became part of the foundation on which Newton was able to build. We shall then consider the notions of space and time as they were conceived in Newton's monumental work, the *Principia*. Newton's mechanics, the topic of that work, must be distinguished from 'Newtonian mechanics' which refers to the later development of that subject. We shall focus on the changing concepts of space, time, and relativity. But it is hoped that the interested reader will pursue the study of the conceptual development of mechanics by using as a guide the annotated reading list at the end of this chapter.

a. *Galileo, the pioneer*

In Greek thought the physical sciences received their start without many of the ingredients that we today consider to be most essential: they were not primarily based on experiments and their quantitative results. To be sure, ancient Greece produced great mathematicians such as Euclid and Archimedes; and it has great scientific accomplishments on its record, such as the measurement of the circumference of the earth to relatively very high accuracy (Eratosthenes) and Archimedes' principle, to mention only two of them. But their view of nature was so heavily intertwined with religious and philosophical views that their science had a very different character from science after Galileo.

[margin note, handwritten:] Greek science based in religion + philosophy, not experiment

Only in the sixteenth century did the first beginnings of science as we know it today begin to dawn. At that time there was in the Western world one highest authority for matters of theology, philosophy, and science: the Church. The science of Aristotle and the astronomy of Ptolemy (Claudius Ptolemy of the second century AD) were considered to be in accord with its teachings and were accepted by it. And there was the prevailing view of the majority of scholars of the time whose scientific opinions were in agreement with the view of the Church. History records that men whose thoughts deviated too far from that accepted view could get into considerable trouble. There was a list of forbidden books, the *Index Librorum Prohibitorum*; and there were trials, excommunications, and worse.

It is largely this historical background, this climate of the times, which must be appreciated in order to understand the lives and work of men with revolutionary scientific ideas such as Copernicus, Galileo, Kepler, and Newton.

Consider the motion of bodies on the surface of the earth when they are not being propelled forward by forces as exerted by man or beast or, for that matter, by the wind. What happens to bodies when all forces cease to act on them? Common observations seem to convince us that in such a situation bodies would slow down and eventually come to rest. Pushing a stone along a flat road or over a sheet of ice indicates that it may take different amounts of time for a body to come to rest. But there seems to be little doubt that without applied forces bodies stop moving. This was indeed the general belief also at the time of Aristotle.

Our notion of forces today is not limited to the action of man or beast, or to the results of wind or other obvious forces. We have learned to recognize as forces less obvious actions such as friction or air resistance. That generalization has contributed to the development of a much more powerful theory of motion. The stone that is made to slide on a road or even on a sheet of ice is not free of external forces: friction forces slow it down. If one idealizes and imagines a smoother and smoother surface one finds that in the limit of no friction at all the stone would continue to move and would not slow down at all. Such considerations lead to the remarkable *Law of Inertia*:

A body on which forces have ceased to act keeps moving with the same speed and in the same direction as it had at the instant when these forces ceased.

This law was first clearly understood by Galileo Galilei (1564–1642). He arrived at it by studying the motion of projectiles in a way that was entirely new. He pioneered experimentation and mathematical description of his observations. At one point he wrote about nature: '. . . that great book – I mean the universe – . . . is written in the mathematical language'. His activity was therefore very much in opposition to the Aristotelian tradition. Neither leading contemporary scientists nor the Fathers of the Church appreciated experimental evidence when it led to conclusions in violation of the accepted doctrine. In fact, at the time of his investigations of projectile motion, Galileo was under house arrest by the Inquisition (for his support of the Copernican system in opposition to the Ptolemaic one). He was then in his seventies [5.1].

Galileo's mathematical analysis of projectile motion made him realize that this motion can be thought of as the simultaneous performance of a vertical and a horizontal motion. The vertical motion takes place under the constant (vertical) force of gravity while the horizontal motion is not subject to any (horizontal) forces. It takes place with constant velocity. And furthermore, these two motions can be studied quite separately and can then be superposed mathematically. Such considerations led Galileo to the above law of inertia. It is therefore often referred to as 'Galileo's law of inertia'.

Galileo pioneering science based on experimentation

But this was only one of many fundamental and seminal discoveries that resulted from Galileo's new approach to science based on experimentation, observation, and mathematics. Among his many other discoveries is one more that is of particular significance here. It is the following important observation: the motion of a projectile shot from a gun on a ship is exactly the same (relative to the ship) as the projectile motion shot from the same gun on shore (relative to the shore). This holds true *even when that ship is in motion* relative to the shore; it is only necessary that the motion be uniform [5.2]. This observation was found to hold for motion in general and not only for projectile motion. It led to the *Galilean Principle of Relativity*:

When two observers are in uniform motion relative to one another they will observe the same laws of mechanics.

The fundamental importance of this principle and its generalizations will become apparent later on. At this point this principle and the law of inertia may simply serve as two examples of the rich research results of the fertile mind and pioneering spirit of Galileo Galilei.

b. *Newton, the first great architect of mathematical theory*

The figure of Sir Isaac Newton stands out like a giant in the history of physics and mathematics. As Professor of Mathematics and Natural Philosophy at Cambridge University in England and as President of the prestigious Royal Society for 25 years he dominated the physical sciences (the term 'natural philosophy' in those days meant 'physical sciences') and mathematics of his time, and his authority weighed heavily over other scientists for generations to come. Newton was born in the same year, 1642, in which Galileo died. This coincidence is symbolic because in many respects he carried on where

Galileo left off. His most famous work, *Mathematical Principles of Natural Philosophy and the System of the World*, or in its original Latin *Philosophiae Naturalis Principia Mathematica* was first published in 1687 ([5.3] and Fig. 5.1). It is known simply as *Principia*. It laid the foundations for the branches of mechanics and of gravitation theory in what is today called 'classical' physics. And it is just celebrating its tercentenary.

Newton's *Principia* is justly famous for several reasons. Firstly, it combined the knowledge of *mechanics* of his predecessors into one single theory, adding a large amount of important original material to it; secondly, it applied the theory successfully to the motion of the planets, moons, and comets in the solar system by means of a *universal law of gravitational force* (see Chapter 7 below); and thirdly, it did all this in a systematic, logically deductive way reminiscent of the work of Euclid in geometry: starting with definitions of terms, axioms are stated from which theorems (propositions) are derived.

Newton's mechanics is based on three laws of motion. At present we shall be concerned here only with the first of these. That first law of motion is just Galileo's law of inertia, but slightly reformulated:

> **A body remains in a state of rest or uniform motion in a straight line unless it is compelled to change that state by an applied force.**

At first sight this is a clear statement that can be easily put to a test. But on further thought it is not so clear. In fact, it seems to require some specification: *with respect to what* does that body remain at rest or in uniform motion? When two bodies fall side by side one of the two bodies remains at rest with respect to the other but at the same time it is subject to the force of gravity. Such cases would contradict the stated law of inertia. It follows that the law of inertia cannot be generally valid; it cannot hold relative to *all* reference frames (Section 3c). The law of inertia is consequently meaningless unless it is specified relative to which reference frame (or frames) it is valid.

Newton resolved this problem by distinguishing between absolute and relative motion. Absolute motion is motion with respect to *absolute space*; relative motion is motion relative to some chosen object. Newton regards the latter with some disdain; it is only offered as a matter of practicality but is not in the great scheme of fundamental things. He writes of absolute space as true and mathematical space which always remains 'similar and immovable'. And this is a notion that he introduces near the very beginning of his treatise. Thus, when he comes to his first law of motion no difficulties arise: the first law refers to absolute or true motion.

Now it is necessary to ask how one is to determine whether a body is at rest or in uniform motion with respect to absolute space. Which reference frame

Fig. 5.1. Title page of the first edition of Newton's *Principia*.

PHILOSOPHIÆ
NATURALIS
PRINCIPIA
MATHEMATICA.

Autore *JS. NEWTON,* Trin. Coll. Cantab. Soc. Matheseos
Professore *Lucasiano,* & Societatis Regalis Sodali.

IMPRIMATUR·
S. PEPYS, *Reg. Soc.* PRÆSES.
Julii 5. 1686.

LONDINI,
Jussu *Societatis Regiæ* ac Typis *Josephi Streater.* Prostant Vena-
les apud *Sam. Smith* ad insignia Principis *Walliæ* in Cœmiterio
D. *Pauli,* aliosq; nonnullos Bibliopolas. *Anno* MDCLXXXVII.

is at rest in absolute space? Newton realized that this is a difficult issue. But, as he put it, 'the thing is not altogether desperate'.

However, the answer is given by Newton only near the end of his *Principia* when he writes about his 'System of the World': the center of absolute space is the center of gravity [5.4] of the whole solar system, sun, earth, and all the planets. This statement is one of the very few in his treatise that is introduced as a hypothesis. But it is clear that he considered it to be of the greatest importance.

By means of his 'system of the world' (we note that he says nothing about the fixed stars), Newton also answers the old and heatedly debated issue whether the earth or the sun is in the center of the world. If the sun were infinitely heavy the center of gravity would be in the center of the sun. The center of the world would then coincide with the center of the sun as Copernicus had proposed. While the sun is clearly not infinitely heavy it is much heavier than all the planets together (more than seven hundred times heavier). Therefore the center of gravity of the solar system is very near to the center of the sun, so that the sun is almost but not quite in the center of the solar system (world). The earth which Ptolemy claimed to be in the center is clearly a good distance away from it (about 150 million km or 93 million miles), and it is not at rest but in absolute motion. The acceptance of the Copernican system in Newton's time (well over a century later) was evidently not as much of a heresy as it was in the time of Galileo.

But let us leave the lofty description of the 'system of the world' and return to earth. Here we have a surprise waiting for us. The relatively mundane motion of projectiles which Galileo had studied is of course also correctly described by Newton's mechanics, and so are many other simple phenomena of motion which Newton himself investigated. But in these descriptions no reference is made to absolute space! They are based on *relative* motion. The projectiles shot from guns are studied relative to the gunner and not relative to the center of gravity of the solar system. Newton was of course aware of this, and he pointed out that one uses relative motion for practical purposes. But then there must exist reference frames other than absolute space relative to which the law of inertia is also valid. We find little discussion of those by Newton.

In the *Principia* the introduction of absolute space is accompanied by that of absolute time. Like space, absolute or true time is to be distinguished from relative time. As he points out, it is the latter that people measure using such means as clocks or calendars. We are well acquainted with relative time. Our clocks measure time by 12-hour intervals starting at midnight or at midday (in the United States), or 24-hour intervals starting at midnight (in Europe). Our calendars measure years starting with some specific event of religious significance, be it the birth of Jesus (Christian), the creation of the world (Jewish), or the migration of Mohammed to Medina (Islamic). And the length of the year may be controlled by the sun (solar calendar) or by the moon (lunar calendar). All these are relative times. We never seem to use absolute time.

The introduction of absolute time is therefore a puzzle. Why did Newton introduce it? Unlike absolute space, the law of inertia does not demand it. But there was a good reason for Newton to introduce these absolutes. He must be viewed within the context of his time. Newton was a religious man who felt that his work would contribute to an appreciation of the glory of God. Absolute space is the 'Sensorium' of God who constructed the world that is contained in that space. The laws of nature exist according to a divine plan. The absolute nature of space as well as of time reflect God's universality, His eternity, and His infinity. These were the kind of thoughts that motivated Newton. In fact, Newton started out as a student of theology but was 'side-tracked' into mathematics and science. After the publication of his *Principia*, however, he returned to theology and spent most of his time on that rather than on science.

Returning to Newton's first law of motion, the law of inertia, we now want to point out a case of circular reasoning. The whole theory starts with the laws of motion including that first law. It then proceeds to develop mechanics; after that it is applied to gravitation and successfully accounts for the solar system; and finally one can compute the center of gravity of the solar system in order to find the location of absolute rest. And that is necessary to give meaning to the first law of motion.

It seems that Newton was aware of this vicious circle. He designed various thought experiments that permit a determination of the location of absolute space. One of these, the rotating pail, will interest us later in the context of gravitation theory. However, Newton did not succeed in proposing an empirical determination of either absolute space or of absolute time, even in principle. Nor has anyone else since then.

For Newton absolute space and time were real. The world consists of absolute space that is occupied by various celestial objects, the sun, the planets, moons, comets, etc. The image is that of a room into which furniture has been placed. But there were critics of Newton's view concerning absolute space and time. The most famous of these was Newton's contemporary, the German philosopher Gottfried Wilhelm Leibniz. He argued that space is nothing but the relationship of the location of objects. Therefore no real existence can be attributed to space by itself. Empty space, space without any objects, is a meaningless concept and cannot be ascribed reality. The debate between the views of Leibniz and Newton is represented by an exchange of letters between Leibniz and one of Newton's disciples, Samuel Clarke. It is a most interesting correspondence (see Alexander 1956).

Despite this and other debates on the subject, Newton's absolute space and time were generally accepted for some two hundred years. This was largely a matter of default: most people paid little attention to it. The ingenuous and prolific German mathematician Leonhard Euler who was only 20 when Newton died contributed much to making Newton's work accessible to a wider audience. He cast Newton's calculus and the mathematical formulation of his mechanics into the form which we use today. He also greatly extended his mechanics of fluids. When further elaborations and extensions of his work brought continued confirmation by experiment and observation, Newtonian mechanics was hailed for its great successes. But the <u>foundations of the theory in relation to absolute space and time</u> were not studied. It was a <u>dominance of pragmatism over a concern for questions of meaning and reality</u>.

Many other important developments of Newtonian mechanics and gravitation theory were made in the late eighteenth and early nineteenth century. Such men as Lagrange, Laplace, and Hamilton, who were both mathematicians, theoretical physicists and astronomers, provided much more powerful mathematical techniques than were available to Newton. They made it possible to predict astronomical events with very great precision. But during this whole period of the elaboration and extension of Newton's work absolute space and time seemed to play no role at all. Everyone worked with *relative motion* and no difficulties were encountered. Whether or not there is an absolute space and time became more and more an issue for the philosophers only and less and less one for the scientists. It became quite clear that there was <u>no empirical way to determine absolute motion</u>.

c. *The relativity of Newtonian mechanics*

How was it possible for Newtonian mechanics to succeed without absolute space and absolute time on which it was based? The answer lies in Newton's genius: he had constructed a theory that was able to survive the misconceptions of its creator. Absolute space and time were found to be scientifically meaningless concepts. They have no empirical content: <u>the solar system is moving relative to the center of our galaxy</u>, our galaxy is moving relative to neighboring galaxies, etc. And they have no value as purely theoretical constructs either. The first law of motion can be given meaning without any recourse whatever to these concepts. Relative rather than absolute motion is all that can be observed and all that has ever been used in mechanics.

History has thus decided in favor of the *relativists*, those who criticized Newton on his absolutism. The earliest of these were his contemporaries,

□ could be other way around — we assume otherwise for simplicity!

Bishop George Berkeley, and Gottfried Wilhelm Leibniz whose correspondence with Clarke has already been mentioned (Alexander 1956). These two philosophers were among the very few who were concerned about foundational questions in mechanics that touched on philosophical issues. Such issues were not raised by physicists until the second half of the last century. At that time empiricism and instrumentalism was championed by scientists such as Heinrich Hertz and Ernst Mach in Germany. They were trying to rid the theory of notions that were not closely related to observable quantities. Henri Poincaré, the French mathematician, inspired by the philosopher Immanuel Kant, tried to view the Euclidean nature of space as just a convention in analogy to what is done in pure mathematics (Poincaré 1952).

One person who made a conscientious effort to reinterpret the law of inertia in a relativistic way, and thereby to rid Newton's mechanics of the need for absolute space and absolute time, was the physicist Ludwig Lange, one of the lesser known scientists of his generation. That was only a century ago, two hundred years after the publication of the *Principia*. Lange introduced the extremely useful notion of an *inertial* reference frame.

In order to give meaning to the first law of motion we do not need absolute space and absolute motion but we can understand it entirely in terms of relative motion. It is simply a matter of interpretation; not a single line of the mathematics of Newton's theory needs to be changed. The key to this interpretation of the first law of motion is to observe that the law is obviously false when relative motion is referred to certain reference frames, but that it is correct when referred to other frames. These latter reference frames are called *inertial reference frames*. They are the frames which are not subject to acceleration i.e. on which no forces are acting.

A frictionless horizontal surface (an ideal sheet of ice) is an example of an inertial reference frame for objects sliding on it. A rotating platform (a merry-go-round) is an example of a reference frame that is not an inertial reference frame. Any object on it would behave as if it were experiencing a centrifugal acceleration and would therefore not move with constant velocity even though there are no forces acting on it. Thus there are many inertial as well as many non-inertial reference frames. Any one of the inertial ones is acceptable for the application of Newton's mechanics.

A laboratory fixed on the surface of the earth is an inertial reference frame in a certain approximation: the spin of the earth about its axis and the rotating motion of the earth around the sun must be negligible compared to the other accelerations of the experiment. We see here the important role that approximations play (see Section 1c).

From here on we shall have to distinguish the mechanics based on relative motion from the mechanics based on absolute space and time as originally envisioned by Newton. We shall call it *Newtonian mechanics* and distinguish it from *Newton's mechanics*. It was the former that was used in actuality all along by Newton as well by his successors since absolute space was never actually available as a reference frame. People were simply not fully aware of this situation. But we now understand why they succeeded without having to worry about absolute space and absolute time. Newton's mechanics is based on absolutes, but *Newtonian mechanics is a relativistic theory*.

Let us now go back to Galileo and his Principle of Relativity. It says that two reference frames (observers) that move with constant velocity relative to one another have something very important in common: the laws of motion relative to one of these and the laws of motion relative to the other one are exactly the same. If a ship moves with constant velocity relative to the shore then the experiments carried out on the ship and those that are carried out on shore will lead to exactly the same laws of motion.

We understand now why this is so: if one of these reference frames is an inertial reference frame then the other one is also an inertial reference frame. When one adds a constant velocity to a body that is already moving with some constant velocity the resultant motion will also be one with constant velocity. Thus, there exists a whole *family of inertial reference frames*: any two members of that family move uniformly (with constant velocity) relative to one another. It is clear that this property is 'transitive': if reference frame A moves uniformly relative to reference frame B, and B moves uniformly relative to C, then A moves uniformly relative to C. Furthermore, each uniform motion has an inverse, the one with the same speed in the opposite direction. All this is certainly true in one space dimension. But one can show that it also holds in all three space dimensions (with speeds in any direction) according to Newtonian mechanics. It follows that the set of all uniform motions forms a mathematical group [5.5]. The reason for emphasizing this relation of all inertial frames to the mathematical concept of a group lies in the wealth of information mathematicians have about groups and their properties. This information can be applied by physicists to great advantage.

If a battleship moves uniformly relative to the shore and it is an inertial reference frame then so is the shore. A missile fired from the ship can be studied by observers from both the ship and the shore. The two observers will see different trajectories but they will find that these trajectories are due to the *same* laws of motion. Suppose we 'transform' the description of the missile's motion as seen by the ship to the description as seen on the shore. Then the trajectory will also transform. But the laws of motion will remain unchanged. We conclude that when one transforms the description of motion from one inertial reference frame to another *the laws of motion remain invariant* (Section 3c). These transformations are called Galilean transformations and we can assert that *the laws of Newtonian mechanics are invariant under Galilean transformations*. The set of all these transformations make up the *Galilean transformation group*.

Recalling our earlier and more general discussion about invariance and its relation to symmetry (Section 3c), we can also say that Newtonian mechanics has a certain symmetry property: its laws remain invariant under the Galilean transformation group, i.e. under the set of all transformations that change one inertial reference frame to another.

It is easy to express these Galilean transformations in a simple mathematical way. Consider motion along a straight line. Choose a point on that line as a reference point; call it O (Fig. 5.2(a)). Distances are measured from that point. The point P, for example, is a certain distance away from O, x meters, say. Now consider a point O′ which is also on that line but which is moving to the right with speed v meters per second. Let us measure time from the instant when O′ passed O. After a time of t seconds has gone by, we shall find O′ at the location vt meters to the right of O (Fig. 5.2(b)). The point O′ is a uniformly moving reference point and all motion can be referred to it just as well as to O. Relative to O′ the point P is at a distance of x' meters. Therefore, the description of P as seen by O′ and the description of P as seen by O are related by (Fig. 5.2(c)),

$$x' = x - vt.$$

We have thus found the transformation from the 'O frame' to the 'O′ frame' of reference. The 'observer' O will measure the motion of P in terms of x and t, plotting different locations x at different times t, while the 'observer' O′

Fig. 5.2. The Galilean transformation (see text).

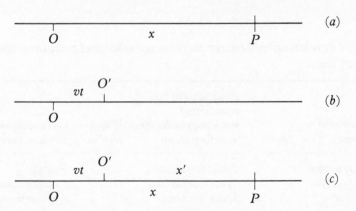

will do the same using x' and t. Both observers use the same clocks: t does not transform, only x does. We can state this formally by

$$t' = t.$$

It may seem pedantic to write down a mathematical statement for such a trivial fact that both observers use the same time, i.e. that time does not 'transform'; but we shall see later how important this matter can be.

This transformation from one reference frame to another when the two frames differ only in that one is moving uniformly relative to the other is called a *boost*. The above equation is the mathematical expression of a boost. Newtonian mechanics is invariant under *boost transformations*.

Invariance under boost transformations is not the only invariance property of the laws of mechanics. There are other, seemingly trivial, invariance properties that will however play a very vital role. Empty space is everywhere the same: it is *homogeneous*. It is also *isotropic*: it is the same in all directions. These trivial facts give rise to invariance properties.

Once it is recognized that there is no preferred point and no preferred direction anywhere in space, so that it is truly homogeneous and isotropic, we have just the properties of a mathematical space, a *Euclidean space*. In such a space a reference point O can be located anywhere. This means that a shift of O to a new location O' which is at rest relative to O will give a new reference point that is just as good as the previous one. Therefore, the laws of mechanics must be invariant under such a shift. These shifts are technically called *translations*. There are of course an infinite number of different shifts possible, namely shifts in all directions and by all different amounts. The resultant family of reference points also forms a 'group' in the above mathematical sense. That's the *translation group*. And we conclude that the laws of mechanics must also be invariant with respect to the translation group.

Table 5.1. *The relationship between properties of space and time, invariance, and conservation laws.*

These properties of space and time	Imply	Invariance of the equations of mechanics under these transformations	Which implies	Conservation laws for these quantities
Homogeneity of time		Time translations		Energy
Homogeneity of space		Space translations		Linear momentum
Isotropy of space		Space rotations		Angular momentum

Finally, the isotropy of space requires that the orientation of the reference frame should not matter. The straight line which we used to derive the formula for the boost transformations can have any orientation whatever in space. That means we can rotate by any angle and about any axis and will obtain another acceptable reference frame. Therefore, the laws of mechanics must also be invariant with respect to *rotations* by any angle and about any axis. These rotations are transformations in three-dimensional space and they also form a group. We thus find invariance also with respect to the *rotation group*.

Summing up we found that Newtonian mechanics is a relativistic theory in the sense that it is invariant under transformations between all different inertial reference frames. These include those that differ from one another only by a translation or by a rotation as well as those that differ by relative uniform motion (boosts). This large group of transformations under which the laws of Newtonian mechanics are invariant characterizes Newtonian mechanics mathematically. It is the *Galilean transformation group* which includes the translation group, the rotation group, and the boost transformation group.

One of the beautiful and powerful results of the use of higher mathematics in the physical sciences is the relationship between the invariance of a theory under a transformation group and the *conservation laws* of that theory. In our case, the invariance of Newtonian mechanics under the groups of translations, rotations, and boosts, can be proven to imply a variety of such conservation laws of which the law of conservation of energy is the best known. For instance, mathematics establishes in this way a completely unexpected and close link between homogeneity and isotropy of space (rather trivial observations) and the laws of conservation of linear momentum and of angular momentum (highly non-trivial laws). This is a staggering accomplishment ([5.6] and Table 5.1).

Annotated reading list for Chapter 5

Alexander, H. G. 1956. *The Leibniz–Clarke Correspondence*. Manchester: University Press. A classic exchange of polemic letters between the philosopher Leibniz who criticized Newton's views on space, time, gravitational force, and the role of God, with Newton's disciple Clarke. Alexander provides a fine introduction.

De Santillana, G. 1955. *The Crime of Galileo*. Chicago: The University Press. A detailed account and commentary of Galileo's problems with the authorities in Rome by a noted philosopher and historian of science.

Holton, G. and Brush, S. G. 1973. *Concepts and Theories in Physical Science*. Reading: Addison-Wesley, 2nd edn. An excellent introduction into the subject with a strong emphasis on its history; the authors are physicists and historians of science.

Jammer, M. 1954. *Concepts of Space*. Cambridge: Harvard University Press. A historian of
   physics presents a fine review of the changing concepts of space from early times to the
   present.
Newton, I. 1687. *Philosophiae Naturalis Principia Mathematica*. Translated by A. Motte and
   revised by F. Cajori 1947: *The Mathematical Principles of Natural Philosophy*. Berkeley:
   University of California Press. This is a difficult book to study even for the experts.
   However, a few non-technical sections such as the 'Scholium' (p.6) and beginning sections
   of 'The System of the World', Hypothesis I (p. 419ff.) are well worth reading.
Poincaré, H. 1952. *Science and Hypothesis*. New York: Dover. This great mathematician was
   also a 'conventionalist' philosopher of science. His views on relative and absolute motion,
   on space and geometry, were of considerable influence. see especially Chapters IV–VII.

6

Imposed consistency: special relativity

a. *Maxwell and the ether hypothesis*

Our twentieth-century technology is to a very large extent based on our scientific knowledge about electricity and magnetism. The important experimental period that led to our understanding of this subject started with the work of Charles Augustin Coulomb in the eighteenth century, though some of this knowledge is much older and some even dates back as far as the ancient Greeks [6.1]. The experimental period continued after Coulomb through the nineteenth century and was a truly international research effort. Of the many contributors some of the best known are (in chronological order): the Italian Alessandro Volta, the Dane Hans Christian Oersted, the Frenchman André Marie Ampère, the German George Simon Ohm, the Englishman Michael Faraday, and the American Joseph Henry. They received many honors for their work. In particular, their names are now used as names of various units in electricity and magnetism.

The most remarkable result of their work was the discovery that the apparently quite unrelated phenomena of electricity and magnetism are actually interconnected in a very intricate way. They are in a sense one and the same phenomenon and they cannot be cleanly separated into purely electric and purely magnetic ones.

Of considerable help in this development was the notion of a *field*. For example, every electric current is surrounded by a *magnetic field*. This means that the space around it is endowed with the ability to exert a force on a magnet (a compass needle for instance). That force is present no matter where that 'test magnet' is placed, although the strength and the direction of that force does vary with location. The force an the test magnet is a measure of that magnetic field. Similarly, every electrically charged object is surrounded by an *electric field*: an electric test charge placed in its vicinity will experience a force.

But not only does an *electric* current produce a *magnetic* field, the inverse is also true: every *magnet* induces an *electric* current in a wire loop that is moving relative to it. The quantitative investigation of this effect led to Faraday's celebrated *law of induction*. There exists therefore a kind of 'intertwining' between electric and magnetic phenomena. In order to express this interconnection as one combined complex of phenomena a new term was created: *electromagnetism*.

Such new insight into nature was of course not gained overnight; it was the result of a long period of experimentation by very many ingenious investigators of whom the above list of names is but a brief sample. Various empirical laws were found and eventually this whole body of knowledge was cast into a precise and quantitative language. Eventually, a set of equations was found from which *all* known laws of electricity and magnetism could be derived by mathematical deduction. This tremendous achievement was the work of the theoretical physicist James Clerk Maxwell (1831–79). His equations became justly famous for unifying into just a few lines such a very large variety of different phenomena and the laws that govern them. This accomplishment is a prime example of unification of scientific knowledge and has already been touched upon in Section 4b. *Maxwell's equations* became the foundations of *the classical theory of electromagnetism*. His 'Treatise on electricity and magnetism' was published in 1873 and played the same role for electromagnetic theory as Newton's *Principia* did for mechanics.

But there was more to that theory than the unification of electricity and magnetism. The theory did not only cast previously known phenomena into a mathematical form. It also made various predictions. One of these was something apparently entirely new: *electromagnetic waves*. These are composed of electric and magnetic fields and contain no matter at all. Furthermore, they propagate in matter-free regions (vacuum) with a speed that occurred in Maxwell's equations as a fixed number, a constant; it was denoted by the letter c. That constant also occurs in various other contexts of the theory. It had to be determined experimentally.

The surprising result was that c, the speed of electromagnetic waves, was related to other known constants and had the same value as the speed of light which had been known for some time and which had been measured repeatedly. Maxwell therefore conjectured that light is nothing else but these predicted electromagnetic waves [6.2]. This conjecture was indeed confirmed in 1880 by the outstanding experimentalist Heinrich Hertz. Later, heat waves, radio waves, and X-rays were found to be electromagnetic waves too. Thus Maxwell's theory not only unified electricity and magnetism with optics (as referred to in Section 4b) but permitted an extension of that unification to include apparently quite unrelated phenomena.

A side remark is necessary at this point. In optics one sometimes speaks of light *rays* rather than light *waves*. But Maxwell's theory deals only with waves and not with rays. There seems to be an inconsistency: rays and waves are after all quite different. This apparent inconsistency is due to the use of the concept of 'ray' in an idealized way. From the point of view of Maxwell's theory it is only an *approximation*: the notion of 'light ray' is used as a very good approximation to 'light wave' in certain circumstances. How a light ray can be an approximation to a light wave can be understood as follows.

Consider a light wave with a flat front. It is somewhat similar to a water wave with a straight front that is produced by a very long floating stick that is being moved up and down. When it encounters a barrier with a hole in it, only part of that wave will proceed through it. That part will move beyond the barrier with exactly the width of the hole. A screen illuminated by it will show a bright spot (the cross-section of the ray); but that spot will not have a sharp edge. Because of its wave nature the ray has a 'diffuse' surface consisting of complicated interference patterns. The image on the screen has an edge of light and dark interference fringes making for a very narrow transition band from the bright spot into the dark shadow. That narrow band is only a few wave lengths wide. The smaller the wave length compared to the size of the hole the narrower is that band of fringes and the closer does the light *wave* approximate a light *ray* with a sharp boundary.

There is a physical connection between an electric charge and the radiation that it emits: whenever an electrically charged object (for example a charged elementary particle like an electron or a proton) is accelerated it will produce electromagnetic radiation. That radiation consists of electromagnetic waves of various frequencies. Which frequencies occur depends predominantly on the acceleration. In general, different frequencies will be emitted with different intensities

The graph that shows the intensity for each frequency is called the *spectrum* of the radiation. One can pick out just one particular frequency somewhat like one chooses a particular radio station from all the stations on a radio dial. If that radiation were visible light, picking out one single frequency would correspond to picking out one single color from the spectrum of the colors of the rainbow. Such radiation is therefore called 'monochromatic'. It has a particular frequency as well as a particular wave length because the frequency determines the wave length and vice versa. The relation between the two is

$$\lambda f = c.$$

Here λ is the *wavelength*, f is the *frequency*, and c is the *speed of light*. If λ is measured in meters and f in vibrations per second then c is necessarily in meters per second. That speed is tremendously high: in a vacuum the number c is 299 792 458 m/s [6.3].

b. *The paradox of the speed of light*

All this impressive success of Maxwell's theory which we have just seen was unfortunately marred by a very serious and very basic conceptual problem. That had to do with the way in which the speed of light occurs in Maxwell's theory. It occurs there as a number, a *constant*, rather than as a quantity that depends on the motion of the light wave.

The speed of an object is a *relative* quantity depending on the frame relative to which it is measured. Elementary reasoning according to Newtonian mechanics requires that if the speed of light is c as measured in a particular reference frame then it cannot also be the same number c relative to a different frame. If the source of the light (a lamp, say) is used as a reference frame then the speed of light should have one value relative to a source at rest and another value relative to a moving source. But that is not the case according to Maxwell's theory.

How can this be? One seems to be drawn to the conclusion that Maxwell's theory is valid only relative to one single reference frame!

Maxwell and his contemporaries were aware of that. They pictured light waves as traveling *on* something. Just as water waves move on water and sound waves move on air, electromagnetic waves move on some medium. That medium was called the *ether*. And c is the speed of light relative to the ether. Therefore, it was argued, Maxwell's theory is strictly valid only relative to the ether. *The ether determines a preferred reference frame.*

It is amusing to observe here a certain parallel between this problem and the problem Newton had in trying to make sense of the law of inertia. He resolved it by postulating absolute space: Newton's mechanics is valid relative to absolute space as the preferred reference frame. And now we find that Maxwell considered his electrodynamics valid only relative to the ether. As is well known, later developments proved both Newton's and Maxwell's notions to be wrong. While their models were incorrect in *that* respect, their equations and the entire mathematical structures of their theories were correct. These theories survived the mistaken notions of their authors successfully. We have already seen how the concept of absolute space had to be abandoned in favor of the concept of inertial reference frames and how it led to a relativistic Newtonian mechanics. The story of the concept of the ether and its eventual demise is part of the quest that finally led to the development of the special theory of relativity.

The problem was therefore to find the motion of the ether because that would be the reference frame for Maxwell's theory. The most reasonable suggestion based on Newton's mechanics was that the ether is at rest in absolute space. But the only way to find that out is to look at the empirical

[margin handwritten notes:] how can c always be the same?

⊗ electromagnetic waves move on ether — velocity c valid only for ether evidence. Numerous observations and experiments were done. Of these we want to present here only three key pieces of evidence. They are in chronological order: the effect of aberration, the experiment by the Frenchman Armand Hippolyte Louis Fizeau, and the experiments by the Americans Albert Abraham Michelson and his collaborator Edward W. Morley.

The effect, called *aberration*, was first observed in 1728 by James Bradley. In order to understand it some preliminary considerations are helpful. Assume it is raining hard and the rain comes down exactly vertically. When I open my umbrella I must therefore hold it vertically to be well protected from the rain. What will happen if I walk very fast? The situation is equivalent to standing still and having the wind drive the rain at the same speed as my walking speed but in the opposite direction. The rain appears to come down at an angle now. Therefore, when I am walking I must tilt my umbrella forward in the direction in which I am moving. The rain is still coming down vertically as seen by someone who is standing still. But it is coming down at an angle when I am walking. That is a matter of reference frame.

Now consider a star that is at rest relative to the sun. When we look at that star from our moving earth we must tilt our telescope very slightly in the direction in which the earth is moving. The star light seems to come from a slightly different direction because we are moving. That is the effect of aberration [6.4].

As the earth goes around the sun it changes the direction of its velocity. Six months from now when it has completed a half-circle it will move in exactly the opposite direction. The aberration effect will then require us to tilt our telescopes in the opposite direction when looking at the same star. In this way a very accurate determination of that small angle can be made. Since we know the speed of the earth relative to the sun (about 30 km/s), as well as the speed of light, we can predict the aberration angle under the assumption that the ether is at rest relative to the sun. The prediction agrees with the observation within the very small measurement errors. And improved observations over the years have repeatedly confirmed this result.

This measurement thus implies that *the ether is indeed at rest relative to the sun* (or very nearly so) and is therefore also at rest relative to Newton's absolute space (since the sun is very nearly at rest in it). Therefore, our earth is moving through the ether at about 30 km/s.

All this was well known in Maxwell's time. It lent support to the belief in an ether at rest in absolute space. But let us now consider another piece of evidence. In 1851 Fizeau measured the speed of light as it moves through water. His experimental set-up permitted this measurement to be made in three different situations: when the water was at rest, when it flowed in the same direction, and when it flowed in the opposite direction to the light. The results were surprising. When the water flowed in the direction of the light beam the light moved faster; but it moved slower when the water flowed against the light. One had to conclude that the flowing water *drags the ether along* to a certain extent.

ether dragged along by substances

The *Fizeau experiment* would imply that *all* substances drag along the ether since there is nothing special about water, although the amount of drag is expected to depend on certain properties of the substance (its index of refraction). But no such drag was indicated by the aberration effect so that there exists a clear contradiction between the Fizeau experiment and aberration.

The third and most convincing empirical evidence came from a series of precision experiments. These were the *experiments by Michelson and Morley*. If it is really the case that the earth is moving through the ether as indicated by aberration then the speed of light must be different in the direction parallel to the earth's motion and in the direction perpendicular to it. A measurement of that difference would therefore be a welcome confirmation of the ether hypothesis. In 1887 such experiments were indeed begun and were improved and repeated by Michelson and Morley. They required extremely high accuracy and were therefore very difficult to carry out [6.5].

The basic idea of these experiments can easily be understood by an analogy with a river. A swimmer takes a certain time t to cross a slowly flowing river and return. But he can also swim downstream and back along the same distance as the river is wide. That will take slightly longer than the time t. The difference is proportional to $(v/c)^2$ where v is the speed of the river and c is the speed of the swimmer (assumed much larger than v).

In the Michelson–Morley experiments v is the speed of the ether relative to the earth and c is the speed of light. If the earth is traveling through the ether and there is no drag, v is the same as the speed of the earth relative to the sun. If there is complete drag of the ether by the earth then v is zero. The accuracy of the experiment must be better than $(v/c)^2$.

To everyone's great surprise the result was negative: no difference in the speed of light parallel and perpendicular to the earth's motion could be detected. If that result is correct (and it has been confirmed repeatedly) one must conclude that the earth is dragging the ether along *completely*. That conclusion is in clear contradiction to both the aberration effect (which implies no drag) and the Fizeau experiment (which implies partial drag).

ether hypothesis incorrect – contradictions regarding drag

The paradox between the aberration effect and the Michelson–Morley measurements resulted in the realization that the ether hypothesis is simply untenable. The empirical search for a preferred reference frame, the ether, has failed miserably. It was this disappointment as well as the influence of

Einstein's work (Section 6c) that made people reluctantly accept electromagnetic waves as traveling in a vacuum without the aid of an ether or any other medium. The ether hypothesis is simply unnecessary. After centuries of belief in the ether this change constituted a revolution in people's thinking. No wonder that it took some time for it to become accepted generally.

Looking back today at the ether hypothesis one sees that it was one of those human prejudices that were taken for granted: 'waves must move in a medium and cannot move in a vacuum'. There has never been direct empirical evidence for an ether; its physical properties had to be strange indeed: it had to be weightless (or very nearly so), sufficiently tenuous to penetrate all matter through which light can travel, invisible, etc. And assuming all this, one was then led to contradictory empirical results.

On the other hand, just doing away with the ether would not have resolved all the problems. The problems of the validity of Maxwell's theory would have remained: with respect to which reference frame is the theory valid?

Now there were countless tests of Maxwell's electrodynamics and they all confirmed his equations to the letter. This means that the theory must be valid relative to the reference frames of the people who made these observations. But these people did not all use the same reference frame! Some were on solid ground and some were on sailing ships; some were in fast moving planes and some in rockets to the moon (though the latter two types of experiments were of course not yet available in Maxwell's time). How can *all* these experiments agree with a theory that gives the *same* speed of light relative to all these reference frames? It is at this point that the special theory of relativity enters the scene.

[margin note: how can Maxwell's theory be right for diff. frames? answer: special relativity]

c. *Einstein's fiat*

It is not the intention here to present a history of the development of the special theory of relativity. Nor do we want to explore the circumstances that may have motivated individual scientists to produce brilliant and far-reaching new ideas. The history of science has its important place and there exist excellent expositions of it relating to special relativity (see Miller 1981 and Pais 1982). Our purpose here is a different one. We want to understand the conceptual problems in their most basic form, and how they were resolved. These matters are not always very clear at the time when feverish research activity, often based on trial and error, tries to clear the fog. They can be appreciated much better with hindsight.

At the beginning of our present century, two great and beautiful theories were known in the physical sciences, Newtonian mechanics and Maxwell's electrodynamics. Both of them have unified countless physical phenomena and have reduced each of them to a few lines of mathematics together with a

[margin note: in contradiction: addition of two velocities]

certain conceptual framework. A host of applications of these theories to specific cases were carried out successfully by means of mathematical derivations from these theories. They have been extraordinarily successful in their predictions; they have both been confirmed on innumerable occasions, and yet these two theories were *conceptually* in contradiction with one another!

This was the basic dilemma of the time. It occupied the best minds in the physical sciences. In its barest outline and with hindsight, the conceptual contradiction that the two theories presented, although it was not clear at the time, was twofold. There was a paradox which was primarily physical and another one which was of a more mathematical nature.

From the physical point of view the simple *law of the addition of two velocities* which is at the very basis of Newtonian mechanics does not hold when applied to the speed of light: let a fast and a slow body move in the same direction with speeds u and v, respectively, relative to some reference frame. According to mechanics (in fact it seems to be common sense) the fast body will then move with a speed $u - v$ relative to the slow body (Fig. 6.1). But now consider exactly the same physical situation but with the fast body (speed u) replaced by a light ray. The front of the light ray moves with speed c. Then the motion of the light ray relative to the slow body is not $c - v$, i.e. it is *not* correctly described by taking the difference between the two speeds. Electromagnetic radiation in a vacuum (and light is a special case of that) is *always* found to move with the *same* velocity c. It is as if the difference between c and v were not $c - v$ but were equal to c. The law for adding or subtracting speeds when the speed of light is involved is somehow not ordinary addition and subtraction.

From the mathematical point of view the two theories, Newtonian mechanics and electrodynamics, *differ in their invariance properties*. In Section 5c we saw that Newtonian mechanics is Galilean invariant. This means that its fundamental equations do not change when one transforms the positions

Fig. 6.1. A surprising result from experiments with light rays. A bicycle and a car move with speeds v and u, respectively, relative to a house. The speed of the car relative to the bicycle is, therefore, $u - v$. But if the car is replaced by a light ray of speed c relative to the house, experiments imply that its speed relative to the bicycle is also c and not $c - v$.

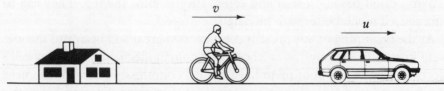

and velocities of objects as seen relative to one inertial reference frame to the relative positions and velocities as seen by another such frame. (For frames in uniform motion relative to one another these transformations are the Galilean boost transformations of Section 5c.) Another way of saying this is that two inertial observers find that their experiments are correctly predicted by exactly the same theory of mechanics because that transformation does not change the fundamental equations of that theory (they remain invariant). But that is just the principle of relativity for mechanics: one inertial reference frame is as good as another. The laws of mechanics are exactly the same for both of them. However, the laws of electrodynamics do *not* remain unchanged under boost transformations because *Maxwell's electrodynamics is not Galilean invariant*! This is consistent with the view that this theory holds only relative to one preferred reference frame. But it is inconsistent with the empirical evidence that indicates a very much greater generality. In any case, if mechanics *and* electrodynamics hold relative to one reference frame, they cannot both hold relative to another one which is moving uniformly relative to it (related to it by a Galilean boost transformation).

The conclusion from this doubleheaded paradox is clearly that one cannot accept both theories as correct. Since they contradict one another, one of them must be wrong and must be given up. But which one? That's the dilemma.

Would it not be reasonable to expect the principle of relativity to hold for both, for mechanics and also for electrodynamics? Perhaps the first person who thought so was the great French mathematician Henri Poincaré. He was of course aware of the difficulty that in this case Maxwell's theory would either have to be discarded or would at least have to be suitably modified.

It was just around this time that two physicists independently made a very odd suggestion. They were the Dutchman Hendrik Antoon Lorentz (1892) and the Scotchman George Francis FitzGerald (1893). They suggested that one could explain the result of the Michelson–Morley experiment if one were to assume that the lengths of all moving objects contract in the direction of their motion. This contraction would have to depend on the speed and, in fact, must take place exactly by a factor of

$$\sqrt{[1 - (v/c)^2]}.$$

Lorentz and FitzGerald showed that if this is assumed to be the case then a measurement of the speed of light parallel and perpendicular to the earth's motion will always result in the same value, in agreement with the Michelson–Morley experiment.

Now this is clearly a lot for someone to believe. Why should things contract like that? Why should a stick of length l change its length to $\sqrt{[1 - (v/c)^2]}l$

when it is set in motion with speed v? Does it have to do with the way bodies move through the ether? It sounds like a very *ad hoc* explanation. To nobody's surprise this *contraction hypothesis* was not considered to be a credible explanation of the Michelson–Morley result; it was not taken seriously.

But in 1904 Lorentz went a step further. He demonstrated mathematically that the Maxwell equations, the fundamental equations of electrodynamics, are not changed (i.e. they remain invariant) when one transforms space *and time* in a very special way. That transformation would replace the Galilean boost transformation. It has since been called the 'Lorentz transformation'. The Lorentz–FitzGerald contraction suggested some ten years earlier entered as an underlying hypothesis into the demonstration. But what would be the meaning of such a transformation? And what sense is there in a transformation of time in addition to one of lengths?

At that point, however, Poincaré realized that he could apply the Principle of Relativity to electrodynamics provided it is interpreted to mean 'invariance under Lorentz boost transformations' rather than 'invariance under Galilean boost transformations'. Of course, one would then have to give up or at least modify Newtonian mechanics because it is not invariant under Lorentz transformations. For some reason the equivalence of the inertial frames is then not given by the Galilean transformations but by the Lorentz transformations (which are the Lorentz boost transformations augmented by the translations and rotations).

The solution to this confused situation was finally given by Albert Einstein (1879–1955). He was motivated by two quite separate matters. One of these was of course the problem of the ether as a preferred reference frame and the contradictory empirical evidence about that. But the other came from his thoughts on Maxwell's electrodynamics and, in particular, on Faraday's law of induction.

We recall that this law refers to the relative motion of a wire loop and a magnet (see Section 5a). That motion together with the magnetic field of the magnet results in a current flowing in the loop. The detailed explanation of this effect as given at the time was not symmetrical. It was not the same when looked at from the reference frame of the magnet and from the reference frame of the loop. Einstein felt that this phenomenon should be exactly symmetrical since only relative motion is involved.

He resolved both difficulties by fiat: he *postulated* that matters should be the way he felt they ought to be, and he deduced from these postulates the necessary consequences. The fundamental scientific paper on that new theory, which became known as the *Special Theory of Relativity*, was

published by Einstein in his 'annus mirabilis', 1905 [6.6]. He based his
theory on two principles which he postulated:
(1) The Principle of Relativity;
(2) The Principle of the Constancy of the Speed of Light.

By the first principle, the Principle of Relativity, Einstein meant a
generalization of its previous meaning. It used to refer to mechanics only.
Einstein meant it to refer to all the laws of physics, specifically mechanics, as
well as electrodynamics. He insisted that they be both valid for *all* inertial
observers. Faraday's law of induction is just one specific instance of the
functioning of this principle.

The second principle stated what he believed nature was trying to tell us all
along through the contradictory empirical evidence from the search for an
ether. With this principle the ether hypothesis was cast aside. The fact that
the speed of light is always the same independent of the motion of the source
of the light was for him not something that needed to be explained. It was a
new law of nature which simply had to be accepted.

His solution was brilliant in at least two respects. He resolved the problems
as he saw them, and he formulated his new theory in a deductive way based
on just these two principles. He demonstrated that Maxwell's electrodynamics is entirely consistent with these principles, and he then constructed a new mechanics (necessarily different from Newton's!) to conform
with them. Einstein therefore chose between mechanics and electrodynamics
in favor of electrodynamics: Maxwell's theory is correct but Newton's must
be replaced.

One fundamentally important deduction from his two principles was how
one inertial frame must be related to another one. He found that they must
be related by the Lorentz boost transformations. No contraction of lengths
was assumed, but the very same contraction that was suggested *ad hoc* by
Lorentz and by FitzGerald now emerged as a consequence. We shall return
to its meaning in the next section.

The above two principles thus *define* how inertial frames must be related.
The Galilean boost transformation is therefore not correct. One must use the
Lorentz boost transformation. Newtonian mechanics must be modified to
comply with Lorentz invariance.

Let us now consider the Lorentz boost transformations in more detail.
Two reference frames R and R' are in uniform relative motion with speed v.
Let us assume that at some instant of time these two frames pass one another
at which time they coincide: the origins of their coordinates are on top of one
another. In both frames time is measured from that instant on. Let these time
measurements be t and t' respectively. Call the direction in which R' moves

relative to R the x-direction, and label all distances along it by x, as measured from the origin of R, and by x', as measured from the origin of R', Fig. 6.2. If an object is seen by hunter R to be located in position x at time t then it will be seen by hunter R' to be located in position x' at time t'. The relation between x and t on the one hand, and x' and t' on the other, is deducible from Einstein's two postulates. It is just the boost transformation that Lorentz had found and that would leave invariant Maxwell's equations:

$$x' = \gamma (x - vt)$$
$$t' = \gamma (t - vx/c^2).$$

The Greek letter γ is by convention used as an abbreviation for the quantity

$$\gamma = \frac{1}{\sqrt{[1 - (v/c)^2]}}.$$

It is just the reciprocal of the Lorentz–FitzGerald contraction factor.

The relationship between the two reference frames, R and R', is of course completely symmetrical: the observer on R sees R' move with speed v, and the observer on R' sees R move in the opposite direction with the same speed. It follows that one should be able to write the above Lorentz transformation also for a transformation from R' to R rather than from R to R'. It should then look exactly the same but with x and x' interchanged, and with t and t'

Fig. 6.2. Clocks in relative motion disagree. The hunter R' is running with constant speed v along a straight and narrow path, the x-direction. As he passes the standing hunter R, they set their stop watches to 0. A rabbit crosses their path and is seen by both hunters. R sees it at distance x when his clock shows time t. R' sees it at x' when his clock shows t'. If they use perfect clocks, we expect $t = t'$. But at the very high relative speed of the two hunters (close to the speed of light), even when one takes into account the time light takes to go from the rabbit to the hunters, there is a noticeable difference; R and R' do not agree on the time when the rabbit crosses. The Galilean equality of time, $t = t'$, does not hold in special relativity.

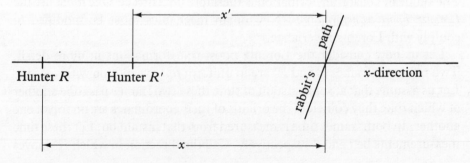

interchanged; in addition, because of the change in direction of the relative speed, all v should be replaced by $-v$. Thus

$$x = \gamma \, (x' + vt')$$
$$t = \gamma \, (t' + vx'/c^2)$$

is the transformation from R' to R. As a consistency check one can substitute these x and t in the previous Lorentz transformation and verify that one obtains an identity.

For the sake of completeness the Lorentz transformations above must be amended by the remark that motion in the x-direction does not in any way change the relative positions in directions perpendicular to it. Thus, if we were to designate the two directions perpendicular to the x-direction as y-direction and as z-direction then $y' = y$ and $z' = z$. This completes the Lorentz transformation.

These then are the Lorentz boost transformations between two inertial reference frames. These transformations do not leave invariant the fundamental equations of Newtonian mechanics. Therefore, according to Einstein's two principles, that theory must be given up. But how can it be that a theory so well confirmed and successful as Newtonian mechanics is now claimed to be wrong? This is an extremely important question and we shall now proceed to answer it.

To this end let us compare the Lorentz boost transformations with the Galilean boost transformations of Section 5c:

$$x' = x - vt,$$
$$t' = t.$$

We see that there are certain similarities between them. If in the Lorentz transformation the factor γ were just equal to 1, and if the vx/c^2 term were absent, the two transformations would be exactly the same. Now let us assume that the speed v of relative motion of the two frames R and R' is very small compared to c, the speed of light. In fact, we need to assume only that the *square* of this ratio is very small:

$$(v/c)^2 \ll 1.$$

Under this condition γ is very close to 1 and the term vx/c^2 is very small compared to t. Therefore, in the approximation where we neglect the very small terms, the Lorentz transformation and the Galilean transformation are the same. In that case there is no distinction between inertial frames as related by Galilean boosts and as related by Lorentz boosts! The equations of the two theories therefore become the same provided only that $(v/c)^2 \ll 1$.

The answer to the question about the demise of Newtonian mechanics in view of the special theory of relativity is therefore as follows. The special theory of relativity does not make <u>Newtonian mechanics</u> wrong or obsolete. All it does is to label it <u>an *approximate theory*</u>. It is approximate in the sense that it is <u>valid only for relative speeds v that are very small compared to the speed of light c</u>. But that is not much of a limitation at all: what speed is not small compared to the speed of light, 300 000 km/s? Our fastest trains are about one million times slower, a rocket that escapes the earth's gravitational pull is about 30 000 times slower, and even the speed of the earth relative to the sun is 10 000 times slower. There is no way in which we can accelerate objects anywhere near to the speed of light. Only by means of the big particle accelerators used by particle physicists is one able to accelerate particles to speeds very close to the speed of light; relativity theory therefore plays an important role only in those instances. In such cases, Newtonian mechanics can indeed not be used any longer. But these are very special cases. Understandably, such speeds for anything but light itself were unknown until quite recently and certainly were not encountered for well over two centuries when Newtonian mechanics ruled supreme. We can conclude therefore that it can continue to rule supreme for *all the phenomena we usually encounter* with the only exception of electromagnetic waves themselves.

Newtonian mechanics usually fine

d. *Simultaneity*

We have seen that Newtonian mechanics is modified by special relativity. This modification is observable only at extremely high speeds, speeds close to the speed of light, speeds never encountered in our daily lives. Therefore, the world of special relativity is an entirely new world. When we enter it we must be prepared for surprises, for we enter into a world in which we have no previous experience. Our intuition, honed on our everyday surroundings, fails. There are in store for us phenomena never encountered before, phenomena that appear strange and unfamiliar, and indeed even phenomena that seem to insult our sensibilities, our 'common sense'.

On the most basic level the surprises start when we have to revise our concepts of space and time or, relativistically, of distance and time intervals. Newtonian physics uses the common belief that a distance between two points, for example the distance between the ends of a meter stick, always measures the same whether the stick is moving or not. Special relativity tells us that this is not exactly true: moving meter sticks do measure less; they show the Lorentz–FitzGerald contraction. Of course, an observer who moves along with the stick would see no such contraction. At ordinary speeds that contraction is so small that it cannot be observed; but at speeds close to those of light this contraction can be appreciable. We shall deduce this contraction from the Lorentz transformation.

A similar revision of our seemingly well-established concepts occurs for time intervals. Here special relativity tells us that a moving clock is slower; it shows longer time intervals and falls behind in showing time. how this comes about can be understood only after we have a means of assuring *synchronization of clocks* that are some distance apart.

Synchronization involves two things: the two clocks must go at the same *rate*, and they must show the same time *simultaneously*. Let us consider the *concept of simultaneity* first.

Simultaneity seems to be a trivial notion. We use it all the time in daily life and don't encounter any difficulties. Yet, if we have to synchronize two clocks that go at the same rate but that are far apart (on earth and on the moon, say), we realize that problems may arise because of the need to establish communication between the two clocks. Such communication involves a time delay, and such a time delay can cause an error unless it can be taken into account exactly. Problems like that can of course also occur over much shorter distances: shouting takes time because of the finite speed of sound, and it also depends on the atmospheric conditions; sending an electric current impulse takes time since it, too, has a finite speed. Reliable communication with known time delay *is*, however, available. According to the second of the principles assumed in special relativity, light has a fixed speed c that is universally the same and is therefore reliable [6.7]. It is most reasonable, therefore, to adopt light signals for the purpose of such communication. As we shall see, the modifications of our conventional concepts brought about by special relativity are largely due to the adoption of the principle of the constancy of the speed of light. Communication by means of light signals (the only communication that is reliable universally) exhibits explicitly how this principle bears on our concepts of space and time.

We are now ready to establish simultaneity between two distant clocks. Let us call A the location of one clock (a clock right next to us), and let B be the location of the distant clock (on the moon, say). We assume that these two locations are at rest relative to one another and also at rest in an inertial reference frame. If we send a light signal to the moon and it is immediately returned to us (by a mirror, say) then it will arrive back at our location A after some time which we shall call t. That time is twice the distance d from A to B (from us to the moon) divided by the speed of light c,

$$t = 2d/c$$

because at a speed c m/s it takes $2d/c$ seconds to traverse a distance of $2d$ meters.

The universality of the speed of light assures us that the speed on the return trip is the same as on the trip out. We have used that fact in the above formula.

We can therefore instruct the astronaut on the moon about the time on our clock at the instant when the light signal is reflected on his mirror. If our clock (at A) shows time t_0 at the instant we send the light signal, the time of reflection of the signal on his mirror (at B) must occur at time $t_0 + t/2$. We have thus established simultaneity between A and B.

This simple procedure can be depicted graphically (Fig. 6.3(a)). We draw the direction to the moon along the x-axis (x-direction) horizontally, and we draw the time axis vertically up. The resulting drawing is called a *space–time coordinate system* or a *space–time diagram*. The clocks A and B are by assumption at rest relative to one another and also with respect to the reference frame in which this coordinate system is drawn. They are depicted as lines parallel to the time axis. Elapsed time is marked at different points on these lines. The time t_0 at which the light signal is sent from A is indicated. That light ray goes from A to B at an angle that depends on the speed (in this case c). It arrives at B at a time that is d/c later than t_0. The light then returns from B to A making the same angle with the time axis but in opposite direction. As is now quite evident from the diagram, the arrival time back at A is $t_0 + t$ if t is the time it took for the round trip. And the arrival time at B is $t_0 + t/2$.

Fig. 6.3. Simultaneity is relative. Two reference frames R (a) and R' (b) observe the same simple physical process: a light ray sent at time t from A is reflected by a mirror B and returns to A. Relative to frame R, A and B are at rest and the round trip takes time t. The point B_1 is *simultaneous* with A_1 because at the instant $t_0 + t/2$ both are halfway in time between sending and receiving at A. Relative to R', however, A and B are moving and are seen as A' and B'; the round trip takes time t'. The point B'_1 is *not simultaneous* with A'_1 (at instant $t'_0 + t'/2$) but with A'_3.

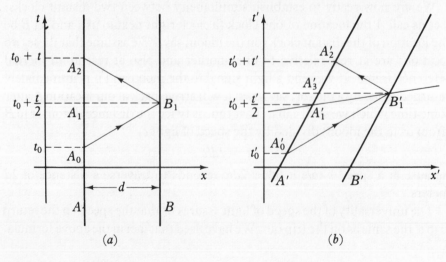

(a)   (b)

Consider now what this procedure for establishing simultaneity looks like as seen by someone in motion. Suppose a spaceship, R', is in uniform motion traveling in a straight line from the moon to the earth (straight at least during the short time of this procedure). An observer on R' will then see both clocks, A and B, move with a certain speed v as shown in Fig. 6.3(b). All quantities as seen by him are marked by a prime: the points A_0, A_1, etc., will be seen by him as the points A_0', A_1', etc. But the light signals from A' to B' and back to A' will have the *same* speed relative to R' as they have relative to R, the previous reference frame. This follows from the second principle. The result is the drawing shown in Fig. 6.3(b).

to observer in motion, points change but light signals same (constancy of speed of light)

But as is evident from the figure, it takes more time for the signal to go from A' to B' than it does from B' to A' as seen by the observer on R'. Therefore the observer on R' concludes that the above procedure did *not* result in establishing simultaneity. He finds that the instant A_1' that corresponds to A (halfway between A_0' and A_2') is *earlier* than the instant B_1'. What is simultaneous in reference frame R is *not* simultaneous in frame R' moving relative to R. *Simultaneity depends on the reference frame*.

This remarkable conclusion lies at the very basis of the special theory of relativity. Only observers at rest with one another will agree on the simultaneity of two events. *though they do agree on space-time coincidences*

In this last statement we have used 'event' as a technical term. It means 'a point in space and a point in time together'. The eruption of a volcano is specified as an event when not only the location of the volcano is given but also the exact time of the eruption. In Fig. 6.3(a) and (b) the points A_0, A_1, A_0', etc., are all events. We shall find this to be a very useful term.

Once we know how to establish simultaneity of two events it is easy to arrange for two distant clocks at relative rest to go at the same rate. We simply produce simultaneity for two consecutive instances. These will give us two equal time intervals which are then assigned the same number of seconds. Clocks that go at the same rates and that also show the same time simultaneously are synchronous. This concludes the solution of our problem of synchronization of distant clocks at rest to one another.

e. *Fast moving clocks and meter sticks*

We are now ready to compare two identical clocks when they are set in motion relative to one another. We shall see that when one looks at a moving clock, it is found to go slower than when it is at rest. The amount of slowdown depends on the relative speed.

The process we used to synchronize two distant clocks can also be used to build a simple *standard clock*. We take two mirrors A and B a short distance d apart. A light signal bounces back and forth between the two mirrors. Each round trip takes the same time *t* which plays the role of a time unit just as the second, minute, or hour does conventionally. The number of round trips measures time. The unit of time on such a clock is thus $t = 2d/c$; it is uniquely determined by the distance *d*. While this is surely a rather unusual clock, it has the advantage of being in principle infinitely accurate and very easy to reproduce. We shall use this standard clock in the following considerations.

Suppose that we have such a clock at rest in a reference frame R'. This frame is moving with constant speed *v* relative to another frame *R* in a direction *perpendicular* to the direction *AB* of the clock (Fig. 6.4). The observer at rest in R' sees this standard clock at rest and reads the standard time: $t' = 2d/c$. On the other hand, the observer in *R* sees the clock moving as shown in Fig. 6.4. He sees the light signal move from *A* to *B* while the clock is in motion. It arrives at *B* after having traversed a distance *D* larger than *d*. As seen by *R* a round trip of the signal takes a time $t = 2D/c$. The ratio of the two times, that seen at *R* (*t*) and that seen at R' (*t'*), is D/d,

$$t/t' = D/d.$$

A simple calculation [6.8] shows that

$$t/t' = 1/\sqrt{[1 - (v/c)^2]} = \gamma.$$

Fig. 6.4. The moving clock is slower. The observer in *e* watches the 'clock' R' move with speed *v* (upward). The clock *R* gets its time intervals from a light ray that goes back and forth between two mirrors *A* and *B* a distance *d* apart. The corresponding light ray in R' is seen by the observer in *R* to travel from A'_0 to B'_1 and from there to A'_2. It travels a distance 2*D* on each round trip while the light ray of clock *R* travels only a distance 2*d*. Since both rays have the same speed *c*, the light ray of the moving clock R' will take longer and that clock will therefore go slower than the clock *R* at rest.

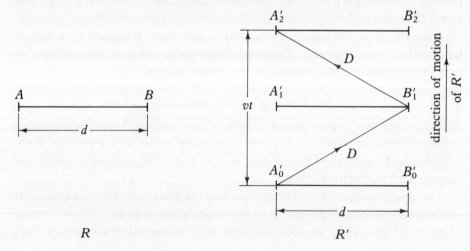

The Greek letter γ denotes the same expression as in the Lorentz transformation (Section 6c). Since this factor γ is larger than 1 for all speeds v that are not zero, the time interval t seen on a moving clock is indeed longer than the same time interval t' measured on a clock at rest. *The moving clock goes slower.*

This increase of time intervals as seen on moving clocks compared to clocks at rest is called *time dilatation*.

The same conclusion as above can also be drawn from the Lorentz transformation, Section 6c. The Lorentz transformation of time intervals is $t = \gamma(t' + vx'/c^2)$. Here the transformation from R' to R is used for convenience. Let the position x' of the clock in R' be at the origin, $x' = 0$. Then this formula yields immediately the desired result $t = \gamma\, t'$.

All this sounds very academic and one wonders what it has to do with the real world. Does nature really behave that way? Can one see this difference in clock rates by a suitable experiment?

One can indeed, but one needs very fast moving objects so that the factor γ is measurably different from 1. An experiment can be made with a particle called a muon. Such a particle can be found in the cosmic radiation that bombards our earth steadily. It can also be produced in high-energy particle accelerators [6.9]. In any case, it is unstable having a lifetime of about 2 microseconds, after which time it decays into other particles [6.10]. Typically, it has a speed very close to the speed of light, e.g. 99% of it. If $v/c = 0.99$ then $\gamma = 7.09$. Therefore, if a muon at rest decays in 2 microseconds, a muon at speeds $v = 0.99c$ is predicted to decay in 7.09×2 or 14.18 microseconds. And this is indeed observed! Time dilatation as predicted by the special theory of relativity can in this way be confirmed by experiment.

muons

There is a very curious phenomenon resulting from time dilatation. It is an apparent paradox that has initiated a great deal of discussion in the early days of the theory: the *twin paradox*. Twins, Art and Bill, decided to test relativity as follows: Art stayed behind in his inertial reference frame while Bill took off with high (but constant) speed v. After he reached his far away planet, he immediately turned around and returned back home to his brother with the same speed v. According to special relativity the reunited twins are now *no longer the same age*: the traveler, Bill, is younger!

twin paradox

There is first the surprise about this result. But then there arises what seems to be a paradox which casts doubt on this result. Special relativity stresses the equivalence of uniformly moving frames; when seen from Bill's frame, the other brother, Art who stayed home, first moved away and then came back. How can one brother become younger than the other *when viewed in the same frame* since the situation seems quite symmetrical?

but to both twins, the other is younger!

The best way to resolve the paradox is to resort to geometry: let us draw what happened in a space–time diagram, Fig. 6.5(*a*). Art, being at rest, appears as a straight line parallel to the time axis. But Bill who first moved to the right and then to the left is drawn as a broken line of two equal lengths. The essential *qualitative* difference is at once apparent: Art remains all the time in the same inertial reference frame while <u>Bill *has to* switch frames in order to return home. The situation is therefore not symmetrical and there is no logical paradox.</u>

But there still remains the question why there should be any difference in ages at all. The time dilatation which some people were willing to accept with the understanding that it is the result of some funny way of measuring time suddenly emerges as *real*. The biological clock behaves the same way as the sophisticated standard clock that we made out of mirrors and light rays! The ratio of the ages of the two twins is exactly given by that factor γ.

In order to make things quite definite, let us assume a relative speed $v = 0.8c$, $\frac{4}{5}$ the speed of light. Then $\gamma = \frac{5}{3}$. If we assume that Bill's clock showed that the roundtrip took six years then Art will claim that it took

Fig. 6.5. The twin 'paradox'. Twin A (Art) stays home while twin B (Bill) travels with speed $4c/5$ to the right. After three years on his clock, B turns around and returns home with the same speed, arriving there after six years; he finds A to be ten years older. In (*a*) the lines of simultaneity are drawn for A (solid) and for B (dashed). The light signals that are sent every year by A are shown in (*b*) and those sent every year by B are shown in (*c*).

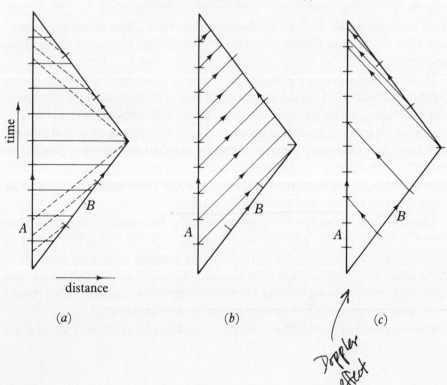

(*a*)   (*b*)   (*c*)

[handwritten margin notes: "Bill switches frames — no paradox b/c not symmetrical"; "Doppler effect"]

$6 \times (\frac{5}{3}) = 10$ years because he claims that Bill's (moving) clock is slow by that factor. The respective years are marked in Fig. 6.5.

One way of comparing their clocks is to draw the lines which connect events that are simultaneous. This is done in (a) of Fig. 6.5. The solid horizontal lines are the simultaneity lines as Art sees them. After five years he is simultaneous with the event where Bill turns around, three years on Bill's clock. The dashed lines indicate simultaneity lines as Bill sees them. As he travels away from Art, Bill's first three years of travel are simultaneous with less than the first two years of Art's clock. When he approaches Art again, his second three years of travel are simultaneous with less than the last two of Art's years (years 9 and 10). But during the very short time while Bill is turning around, he is simultaneous with years numbered three through seven of Art's time. Acceleration changes the simultaneity lines very rapidly.

Another way of comparing the two clocks is to have each brother signal to the other at the end of each year according to his own time. In this way they can keep track of one another's time and see how the age difference arises. The light signals sent by Art to Bill are marked in (b) of Fig. 6.5. As they recede from one another the frequency with which Bill receives the signals from Art is slowed down by a factor of three: Bill receives one every three years only. On the other hand, during the second half of his trip when they are approaching one another, Bill receives three signals every year. This decrease and increase in frequency as the source recedes and approaches, respectively, is known as the *Doppler effect*. It is well known from everyday life: the whistle of an *approaching* train sounds *higher*, of a *receding* train *lower*. At the extremely high speeds we are dealing with these changes are very large. Computation shows that they are higher by a factor of 3 and lower by a factor of $\frac{1}{3}$, respectively.

Since Bill receives the signals with decreased frequency exactly half of the time and with increased frequency the other half of the time, the average frequency of reception of signals is

$$\tfrac{1}{2}(\tfrac{1}{3} + 3) = \tfrac{5}{3},$$

which is exactly the factor γ. Thus, Bill receives ten signals from his brother during his six-year trip and concludes that his brother Art's clock is fast by a factor $\frac{5}{3}$.

Similarly, when Bill signals to Art at the end of each of his six years ((c) of Fig. 6.5), Art receives only three signals by the end of the first nine years because of the lowered frequency (by a factor of $\frac{1}{3}$) during the time Bill was receding from him. On the other hand, during the last year three more signals arrived because of the increased frequency (by a factor of 3). The average frequency of arrival is therefore

$$\tfrac{9}{10} \times \tfrac{1}{3} + \tfrac{1}{10} \times 3 = \tfrac{3}{5},$$

which is just $1/\gamma$. Art therefore receives only six signals from his brother Bill and concludes that Bill's clock is slow by a factor $\tfrac{3}{5}$. Both twins agree that Art got ten years older and Bill only six.

There is nothing wrong with the clocks. Nor do the artificial clocks show a different rate from the biological clocks. What is wrong is our concept of time: it is neither absolute (as Newton thought) nor universal (as we are wont to believe). Time *depends on the history of the traveler through space.* Clocks really do go at different rates. This is not a matter of appearance or of incorrect knowledge. It is not an epistemic matter but an ontic one.

Only clocks at rest will show the same time. Clocks in motion depend on that motion and on the change from one inertial frame to another during that motion. In other terms, one can state that time is not *public* as our 'common sense' dictates but *private.*

The other big surprise that special relativity has in store for us is length contraction. As stated earlier, a meter stick at rest in reference frame R' will measure less than one meter when seen by an observer in R to whom the stick is in motion along the length of the stick. This follows from the Lorentz transformation by an argument quite similar to the one that gave us the time dilation.

Let us look at the length of a stick that is at rest along the x-axis of R'. If one end of it is at the origin of R', the other end will be at $x' = l$ if l is the length of the stick in its rest frame. Being at rest in R' the stick is moving with velocity v relative to R. The observer in R sees the two ends of the moving stick a distance x apart. The two lengths x and x' are related by the Lorentz transformation from R' to R,

$$x' = \gamma\,(x - vt).$$

Since R measures x with its two ends simultaneous, $t = 0$. Therefore, $x' = \gamma x$. The length of the moving stick (seen by R) is $x = x'/\gamma = l/\gamma$. The moving stick is shorter than the stick at rest since $1/\gamma$ is less than 1. This contraction is exactly the prediction made years earlier by Lorentz and FitzGerald.

Unfortunately, it would be very difficult to put this effect to a direct test. There are no one-dimensional objects (like meter sticks) available that move with speeds close to that of light. For three-dimensional objects such as spheres the effect is masked and cannot be seen [6.11]. But there are indirect verifications of this effect because the Lorentz transformation is used in many other contexts which *do* permit experimental confirmation. We shall encounter some of these below (for example, in Note 6.14).

At this point the question may well be asked: 'Do time intervals and sticks *really* change or is this just an "illusion"?' Since the difference in their sizes

[handwritten margin note: time not absolute nor universal (private not public)]

can actually be *measured* there is no doubt that this is a real effect. But, of course, it refers to measurements made relative to *different* frames. A stick does not change its length while moving with a given speed relative to any *one* fixed inertial frame. Nor is this the result of 'a peculiar way in which length and time are defined'. Einstein himself has made this very clear since these were questions that were raised in all seriousness in the early days of the theory.

In addition to time dilatation and length contraction there is one other direct consequence of the Lorentz transformation that is of special interest here; it is necessary for the logical consistency of the theory. When the speed of a light signal is always c then any motion of an observer in the same or in the opposite direction must not change that speed; it cannot subtract or add to it. Speeds do not add or subtract as numbers do. This point was already made earlier. It was necessary in order to account for the Michelson–Morley result.

How speeds 'add' follows easily from the Lorentz transformation. If x is the distance travelled by an object and t is the time interval it took then the speed u of that object is $u = x/t$. Similarly, in reference frame R' the speed would be $u' = x'/t'$. We can therefore simply divide the transformation for x' by the one for t' and find

$$u' = \frac{x'}{t'} = \frac{x - vt}{t - vx/c^2} = \frac{u - v}{1 - vu/c^2}.$$

{ how speeds "add"
} (just like Newtonian except
for denominator)

The numerator of the above fraction is the usual difference between speeds as we would expect it in a situation such as shown in Fig. 6.1. The speed of the car relative to the bicycle (which is R') is expected to be $u' = u - v$. But there is also the denominator of this fraction. That denominator is just 1 when vu/c^2 is very small (i.e. when $v \ll c$); but when this is not the case, the denominator can change the result considerably. In the extreme case when the car is replaced by a light ray (speed c) we find from the above formula by replacing u by c that $u' = c$! The speed of light that is c relative to the house is also c relative to the bicycle.

This is of course not a new result. It is contained in the Principle of the Constancy of the Speed of Light that stands at the head of the theory. And this principle has been incorporated into the above formulae. But it shows how the composition of speeds became modified as a result of assuming that principle. It is a check on internal consistency. It also assures us that the Michelson–Morley result is fully accounted for.

Another application of the above new law for the composition of speeds lies in the explanation of an old experiment. The 1851 Fizeau experiment (Section 6b) found that the speed of light in flowing water differed from that in water at rest. He attributed that to a partial drag of the ether by the water.

It finds its explanation by the special theory of relativity in the peculiar way in which speeds add according to the above formula. That formula is in full accord with the Fizeau results [6.12].

f. *The conversion of matter into energy*

The technical term 'mass' is a measure of the inertia of a body. It tells how hard it is to accelerate the body with a given force. The larger the mass the smaller the resultant acceleration. This is a relation of inverse proportionality. In order to compare the masses of two bodies one can subject them to the same force (for example to a spring under tension) and measure their accelerations. The ratio of their accelerations will be equal to the inverse ratio of their masses, in symbols, $a_1/a_2 = m_2/m_1$. Since accelerations can be easily measured in terms of conventional units (for example in m/s^2 [6.13]), mass ratios can thus be found. Then it remains only to choose a unit of mass to express all masses quantitatively. The metric unit of mass is the kilogram, defined as the mass of a cube of water 10 cm $\times$ 10 cm $\times$ 10 cm. (One gram is the mass of a cube of water 1 cm $\times$ 1 cm $\times$ 1 cm.)

Since the end of the eighteenth century one of the fundamental laws of the physical sciences has been the law of conservation of matter. More precisely, this meant the law of *conservation of mass*. Mass is indestructible. No matter what physical or chemical process may take place, whether a log of wood burns, a piece of sugar dissolves, or an explosion takes place, the total mass of all the matter before the process started is exactly equal to the total mass after it is finished. The state of matter may change, of course; part or all of the system under study may change its phase, solid, liquid, or gaseous. But when the masses of all of its components are measured nothing is gained and nothing is lost.

Well, scientists had a big surprise in store. When the special theory of relativity was developed, one of its unexpected consequences was that this law of indestructibility of matter is false, that matter can be created and destroyed. *Matter can turn into energy and energy can turn into matter*.

In order to understand this better we must examine the other great fundamental law, the *law of conservation of energy*. Energy is one of the most important concepts, and yet it is not easy to explain because it is much more abstract than, say, mass. Of course, the term 'energy' is used widely in everyday language. But there it has a variety of meanings that vary from vitality to power. The technical usage of the term 'energy' is best clarified by a few examples. Lifting a weight requires energy. The energy that is used goes into the work done; 'work' is here the product of the force exerted (which we can assume to be a constant force) times the height by which the weight is raised. The energy may come from muscle power, in which case it is *chemical energy* of processes taking place inside the body. It may come from an electric motor that raises a weight on a platform, in which case it is *electrical energy*. Or it may come from a steam engine, in which case it is *thermal energy*. These examples show the conversion of chemical, electrical, or thermal energy into work. Other forms of energy include *kinetic energy* (energy of motion) and *potential energy*. The latter is energy due to location when a force is present: a brick on a window sill and another on the roof of a house have different potential energy because when they fall the first one will reach the ground with a smaller speed, i.e. smaller kinetic energy, than the latter. Their location matters; the fall converts potential energy (energy of location) into kinetic energy.

Physicists know how to express each of the various forms of energy quantitatively. The law of conservation of energy is a precise quantitative law: in any physical or chemical process the *total* energy (the sum of all forms of energy) does not change. The total energy present after a process is completed is exactly equal to what it was before the process started. The law says nothing about the various forms in which the energy might be present. It allows all changes of these forms of energy during the process.

We can now make the failure of the law of mass conservation more precise. What relativity theory really tells us is that matter is *yet another form of energy*, mass energy. Every amount of mass corresponds to a certain amount of energy. Quantitatively, that is the meaning of the famous relation

$$E = mc^2.$$

Given a mass m one obtains the corresponding amount of mass energy by multiplication by c^2. Once this relation is established [6.14] the law of conservation of energy simply takes over. Mass energy must now be included among all other forms of energy. The law of conservation of mass has disappeared. It existed only as long as that particular energy, mass energy, was not being converted to some other form of energy. If mass energy is conserved, so is mass. It is, as it were, like a law of 'conservation of chemical energy' for processes where no chemical changes are involved.

These predictions have found ample empirical confirmation. While the fundamental conversion of mass energy into other forms of energy occurs on the level of atomic nuclei the existence of chain reactions permits this phenomenon in certain special cases to be amplified to macroscopic sizes. In fact, this can happen in the most spectacular way: confirmation of the relativistic prediction of conversion of matter into energy is found in nuclear power stations, in nuclear bombs, and in the production of energy in the interior of stars such as our sun.

But there is another aspect to the above relation. If mass corresponds to energy, energy corresponds to mass: every form of energy acts like a mass, has inertia, and makes it harder for an applied force to cause acceleration. It must therefore also be possible to convert other forms of energy into mass energy, to convert energy into matter.

All these predictions have indeed been borne out. The conversion of energy into matter has been observed on the elementary particle level (see Section 12a); and one can demonstrate (in the physics of atomic nuclei) that increased potential energy makes things heavier. This remarkable phenomenon can be expressed most poignantly by stating that a wound-up watch weighs more than an unwound one. The increase of the potential energy of the wound-up watch-spring increases its mass and makes it heavier. That potential energy is not in the atomic nuclei but in the binding between the atoms and molecules of the spring metal. It is of course much, much too small to be detected directly.

This remarkable scientific revolution that 'killed' the law of mass conservation raises some most disturbing questions. How was it possible that during all these centuries of scientific explorations mass energy has never been observed to change and to be converted into any other kind of energy? How could scientists have been fooled like that? How can we prevent such a thing from happening again?

Today we know the answers to these questions and they are not difficult at all. Processes in which matter changes into other forms of energy occur both naturally and artificially (initiated by man). The *naturally occurring* ones that take place on the elementary particle level (involving only a few such particles) were obviously not accessible until the availability of very recent technology. We shall return to them in Chapter 12. Those that occur on a massive scale and are therefore macroscopically observable were not recognized as such: the energy source of the sun and the stars was simply not understood. The *artificial production* of energy from matter similarly involves either just a few elementary particles and has not been accessible until recently, or it requires very sophisticated procedures, those involved in nuclear power reactors or in nuclear bombs. Naturally or artificially, macroscopic effects of such conversion are always basically processes involving the atomic nucleus. And that is a very recent field of inquiry.

Concerning the future prevention of instances where scientists are being fooled into believing something that is later found to be false: that can never be assured; it is part of the progress of science. We shall pursue this issue further in Chapter 8.

But let us return to $E = mc^2$. When one computes the amount of energy present in a given mass the result is staggeringly large. If one single gram of a substance were convertible in its entirety into heat energy, one would obtain approximately 25 million kilowatt-hours. That would be the same

amount of energy that is obtained from burning 3000 tons of coal! Unfortunately because of the energy needs of mankind but fortunately because of its
misuse, it is impossible to convert *all* the mass of a substance into energy. In
actuality only a very small fraction of a substance can be so converted. One
gram of pure uranium 235 [6.15] (even enriched uranium contains only a few
percent of uranium 235) would permit only 1 part in 1000 of its mass to be
converted. Nevertheless, the typical nuclear fuel produces by a factor of
millions more energy per kilogram than conventional (chemical) fuel.

The effect on mankind of this discovery, especially in its potential for war,
is known to everyone and needs no further elaboration here. But it is
important to contemplate that the little formula above which provided for
these awesome implications originated in 'highly academic' studies on the
nature of space and time, a subject matter few people outside the scientific
community would consider to be of much interest.

g. *Space–time geometry*

One of Einstein's teachers at the Institute of Technology in Zürich, Switzerland (the Eidgenössische Technische Hochschule), was the mathematician
Hermann Minkowski. Years later, 1907–08, this mathematician made an
extremely important contribution to the theory developed by his former
student. He suggested a geometric representation for relativity so that many
of the strange relations between space and time can be pictured and much can
be understood without the use of algebra.

His idea was very simple: since the Lorentz transformation on which the
special theory of relativity is based involves a transformation of space as well
as of time one may treat time just like another dimension of space, a fourth
dimension, as it were. This very fruitful idea of a four-dimensional 'space',
three dimensions of ordinary space and one time dimension, became known
as *Minkowski space*.

Let us return to the space–time diagrams which we used in Section 6d. Just
as we used a line to describe the location of a clock at different times, we can
also describe the motion of the tip of a light ray in this way. Consider a light
ray that was produced at the origin O of the diagram and moves to the right.
The result is Fig. 6.6 in which the light ray now appears as a straight line at
an angle. For any instant on the t-axis (t-direction) we can draw a line from
that point parallel to the x-direction; where it intersects with the light ray line
we draw a vertical line down to the x-axis (x-direction). The point of
intersection with the x-axis tells us the distance traveled. Thus, for $t = 2$
seconds we find $x = 600\,000$ km. The light ray line is therefore a graph, a
geometric description of the motion of the tip of the light ray. And the angle it
makes with the time axis is determined by the speed c. Space–time diagrams

give not only qualitative but also quantitative information. We only need to provide suitable scales to the axes.

The *algebraic* description of the motion of that same light ray is $x = ct$. This description and the geometric description are completely equivalent. They contain exactly the same information.

But consider now a more general situation: a light source between two parallel horizontal screens very close together so that light can spread out only in the plane between the two screens. We can think of it as a lot of light rays going in all directions from the source but always remaining in the plane. If we follow the tip of any one light ray from the light source out in some arbitrary direction it will reach a distance r km from the source after a time t seconds has elapsed. The formula for its motion is therefore $r = ct$, or $c = r/t$. This algebraic description of the motion can be depicted geometrically as follows. The plane in which the light spreads is taken to be perpendicular to the paper; let's call it the xy-plane because we can draw two perpendicular axes on it, an x-axis and a y-axis. The point in the plane where the two axes cross is again called the 'origin'; it is denoted by O. We locate the light source at O. Now we draw the time axis from this point upward as in Fig. 6.7. The motion of the light rays as they spread out from O now appears as a *cone* with its tip at O. This can be seen in two different ways.

Consider any plane that contains the t-axis as a line. It intersects the cone in two straight lines OA and OB (Fig. 6.8). These lines are the graphs of the motion of two rays. One of these travels in the direction OP in ordinary space (in the xy-plane) and the other in the opposite direction, OQ. They are just like the graphs we considered before. Had we drawn a different plane it

Fig. 6.6. The motion of a light ray in the x-direction.

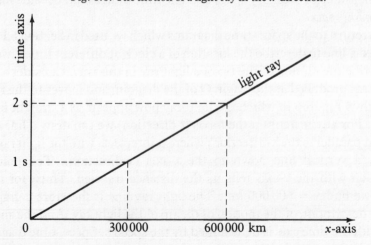

Fig. 6.7. The light cone for a source at O.

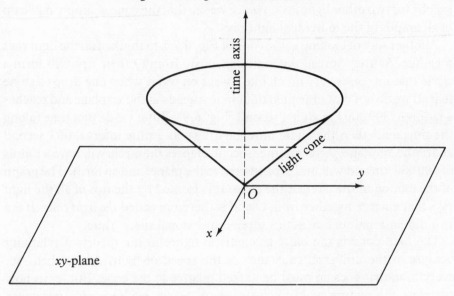

Fig. 6.8. The light cone intersected by a plane containing the time axis.

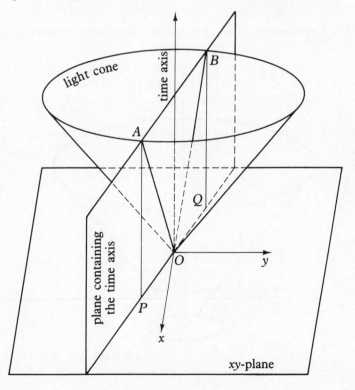

would have intersected the cone in two different straight lines, giving two graphs for two other light rays. Hence we see that the cone is simply made up of all graphs of the individual light rays.

Another way of looking at the cone of Fig. 6.7 is to think of all the light rays together. As they spread out simultaneously from O their tips will form a circle that increases very much like a wave on water when one drops a stone in it. If we do not plot time then this circle spreads in the xy-plane and reaches a radius of 300 000 km after 1 second (Fig. 6.9). But if we do plot time (along the time axis) then the circle must be drawn at a time interval of 1 second above the xy-plane. After another second elapses the circle will have a radius of 600 000 km and will be 2 seconds above the plane, and so forth. The graph of the motion of the circle is the cone. It is formed by the tips of all the light rays that emerge together from O and is therefore called *the light cone*. It is a two-dimensional surface in this three-dimensional space–time.

The light cone is the basic geometrical figure in the theory of relativity because of the universal constancy of the speed of light. As we shall see, everything that goes on must be judged relative to the cone. But let us now complete the picture of Minkowski space. So far we have only *two* space dimensions and one time dimension, and we need *three* space dimensions and one time dimension.

Fig. 6.9. A wave spreading from a point reaches a distance of 300 000 km after one second.

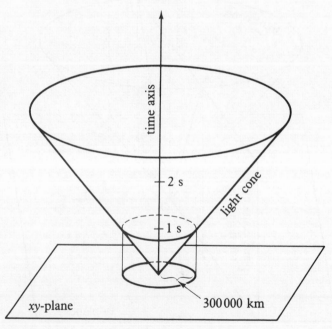

That, surely, poses a serious problem for a draftsman. How can one draw a three-dimensional space in place of the two-dimensional xy-plane perpendicular to the paper? We cannot, of course. But we can imagine it. Mathematicians do this all the time. We can think of the two-dimensional xy-plane as a three-dimensional xyz-space, our ordinary three-dimensional space. In any case, the time axis must remain where it is. Minkowski space as four-dimensional space (xyz-space and the time dimension) must therefore be imagined and cannot be drawn. The best we can do is to draw a plane and think of it as our three-dimensional space, the xyz-space. Our drawing of four-dimensional Minkowski space then looks like Figs. 6.7–6.9. With this in mind a point source located at O sending out light in *all three* space directions is still drawn as before. But now the tips of the rays that emerge from O and spread simultaneously in all directions form a sphere instead of a circle: the circle in Fig. 6.9 represents the *two-dimensional surface* of an expanding light *sphere*, somewhat like the surface of an expanding soap bubble. The light cone shows the *motion* of such a light sphere. The surface of the light cone is *three-dimensional* even though the drawing looks like a two-dimensional surface: two dimensions of the surface of the light sphere and one time dimension.

Minkowski space is a four-dimensional mathematical space that is physically best described as *space–time*. It combines ordinary space and time into one single 'package' which is found to be very convenient for the graphical representation of many features of the special theory of relativity. As we shall see, this unification of space and time goes much deeper than would be expected from the 'package'. We have already seen one indication of that in the twin paradox: the time elapsed for a space traveler (according to his own clock) depends on his past motion. Two travelers who start a trip at the *same* time and who travel along different ways (and possibly also with different speeds) will upon their reunion find that *different* time intervals have elapsed according to their respective clocks: they will have aged by different amounts.

One can think of space–time with three space directions and one time direction also as one single 'space' of *four* directions. To this end one can convert the time direction into a direction along which one measures length just as for ordinary space directions. A time interval t multiplied by the speed of light c gives us *the distance that light travels during the time t*. One can plot such distances ct instead of t along the vertical axis, and it will give us the same information as t gives. In this way one has four directions of lengths in this four-dimensional space. But this space is not just like ordinary space with one more direction because the extra direction which comes from plotting t or ct does not 'combine' with an ordinary space direction in the 'right' way. This is why this 'space–time package' is called 'Minkowski space' rather than 'four-dimensional Euclidean space' [6.16].

The three-dimensional ordinary space (our *xyz*-space) is of course infinite; our symbolic picture of it (the *xy*-plane) is correspondingly also infinite. But so is time. We must draw the time axis not only into the *future* direction (up, on the paper), but also into the past direction (down, on the paper). Similarly, we can also extend the light cone into the *past* as in Fig. 6.10. While the *future light cone* of the previous figures represented the outward expansion of a spherical light pulse, the *past light cone* represents a spherical light pulse that is contracting as time increases until its radius vanishes (at the point O). This latter situation is of course not the description of something that can happen in nature. Nevertheless, the emerging picture of a *double-cone*, a future light cone and a past light cone, will be very useful. This double-cone is generally referred to as *the light cone*. Its algebraic description consists of $r = ct$ for the future light cone, and $r = -ct$ for the past one (t is negative in the past since we have chosen O to have time $t = 0$). These two equations can be combined to $r^2 = c^2t^2$, which describes the double-cone.

Fig. 6.10. The double light cone and three-dimensional space.

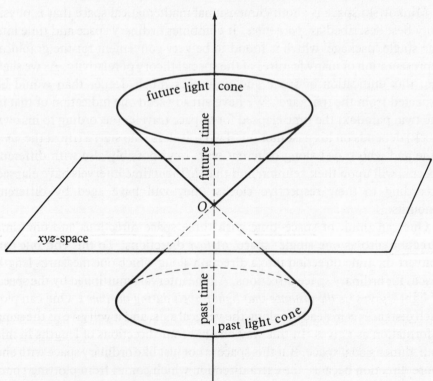

We see already that Minkowski space represents the whole world, past, present, and future. And the light cone (the double-cone) separates this world into regions that have different meanings with respect to time. We shall turn to this aspect next.

Minkowski space permits an easy way of representing all possible motions of a body. It does this by a single line which is called that body's *worldline*. But for this purpose the body must be idealized as a point. In Fig. 6.11 one worldline *AA'* shows an object at rest, and another worldline *BB'* shows one in uniform motion. These are already known to us from Fig. 6.3. A third worldline *CC'* shows a more general motion. In all cases the speed is less than

Fig. 6.11. Worldlines: *AA'* is the worldline of an object at rest. *BB'* is one of an object in uniform motion. Its angle with the time direction is less than that of the light cone. *CC'* is the worldline of a more general motion. At every point *P* on it a tangent line can be drawn; it will have an angle with the time direction less than that of the light cone with the time direction.

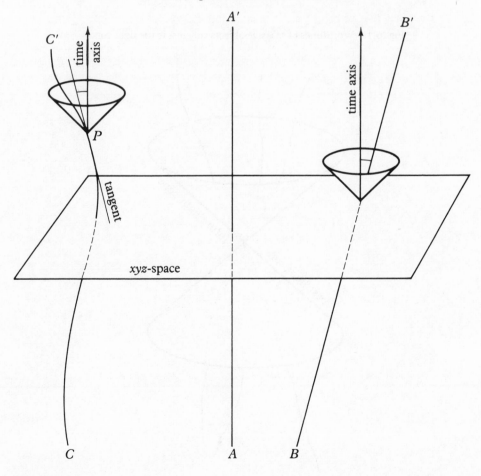

the speed of light c so that the worldline and the time axis (time direction) must at any point make an angle that is *smaller* than the angle which the light cone makes with the axis. The smaller the speed the smaller is the angle.

Now the dynamics of special relativity does not permit acceleration of a massive body to speeds as fast as light. The speed of light is therefore an upper limit that can never be reached [6.17]. It follows that the worldline of *any* uniformly moving massive body must make an angle with the time axis that is less than the angle the light cone makes.

Suppose the worldline CC' is that of a flying bird. A point P on CC' has a unique time t and a unique position r. This means that at time t the bird will have reached position r. At any such point P on that worldline one can draw a straight line tangent to CC' as shown. That straight line tells the velocity of the bird at that time t; that straight line is also the worldline of an object in uniform motion which has a velocity that is exactly the same as the velocity of the bird at that instant.

Fig. 6.12. Worldlines of massive objects stay inside the light cone.

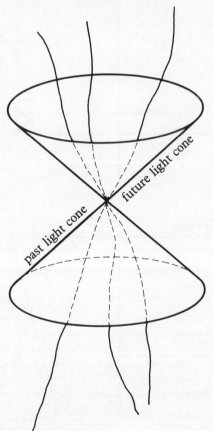

Since the light velocity is the limiting velocity, all worldlines of massive objects must have (at all points) smaller angles with the time axis than the light cone has. Therefore, all worldlines going through the tip of a light cone must be entirely inside that cone (Fig. 6.12). Only light itself has a worldline that is *on* the cone. This leads us to the conclusion that only points inside (rather than outside) the future light cone can be reached by a traveler who starts at the tip of the light cone. Similarly, the tip of the light cone can only be reached by travelers that start *inside* (rather than outside) the past light cone.

A point P in Minkowski space is equivalent to the specification of a particular instant in time *as well as* of a particular point in (ordinary) space. It is therefore exactly what was called an *event* in Section 6d. The above conclusion can now be stated as: *all accessible future events lie inside the future light cone*, and *all events from which the tip of the light cone is accessible lie inside the past light cone*. Of course, if we include in our notion of accessibility also accessibility by means of light rays then the above statement must be amended by replacing 'inside' by 'inside and on' the light cone.

Since the tip of the light cone at the origin O is at time $t = 0$, *all* events in the *xyz*-space through O are at time $t = 0$. They are therefore all *simultaneous*. For that reason the *xyz*-space is often called the *now-space* for any observer at $t = 0$. For such an observer, all events above the *xyz*-space are in the future and all events below are in the past. But the events in the future are of two kinds, those that are accessible and those that are not (Fig. 6.13). All events

Fig. 6.13. The future and the past inside the light cone are accessible. But outside the light cone they are not accessible by any object starting at O.

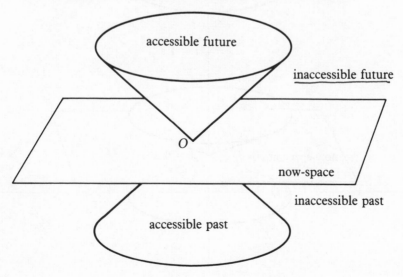

outside the future light cone and above the *xyz*-space are in the future but are inaccessible because one would have to travel from *O* with a speed faster than the speed of light in order to get there *in time*.

Similarly, the origin *O* is inaccessible from any point in the past that is outside the past light cone. Even communication by light signals is not fast enough to get there in time.

At a later time, say at time $t = 2$ seconds, all events that are 'now' are in an *xyz*-space through $t = 2$ which is parallel to the one through $t = 0$. We have a different *now-space*: it is ordinary *xyz*-space two seconds later. If we draw a light cone from that event (at $t = 2$ and at $x = 0$, $y = 0$, and $z = 0$ of *xyz*-space) we find a different region of Minkowski space that is inaccessible (Fig. 6.14). The importance of the light cone is becoming increasingly

Fig. 6.14. At $t = 0$ seconds the now-space is different from the now-space at $t = 2$ seconds.

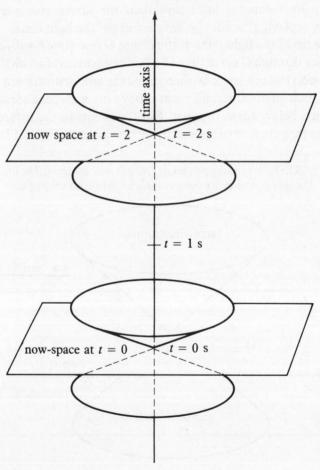

evident. Given an arbitrary event (point in Minkowski space) the light cone with the tip at that point will separate the world into accessible and inaccessible regions. There are inaccessible regions from anywhere one starts. No such restriction exists in the old Newtonian theory.

But what is the meaning of that region of space–time which lies outside the light cone and which is inaccessible from the origin? The answer to this puzzling question is closely related to the concept of simultaneity.

If we recall Fig. 6.3 we remember that simultaneity depends on the reference frame. Two events A_1 and B_1 that are simultaneous relative to the frame R (Fig. 6.3(a)) are not simultaneous relative to R' (A'_1 and B'_1 of Fig. 6.3(b)). The line A_1B_1 that indicates equal time in R is parallel to the x-axis, but that same line as seen from the moving frame R', namely $A'_1B'_1$ is *not* parallel to the x'-axis. Tilted lines like that indicate simultaneity in a moving frame.

If we replace the x-axis of Fig. 6.3 by three-dimensional ordinary space we obtain Minkowski space–time. That tilted line $A'_1B'_1$ that indicates simultaneity in a moving frame then becomes a tilted xyz-space in our drawing (Fig. 6.15). It indicates that the now-space for a moving observer in R' is 'tilted' relative to the now-space of the observer at rest in R. More generally, one can draw a tilted xyz-space that passes through any given event outside the light cone and through the origin. That tilted xyz-space is the now-space for *some* moving reference frame.

Fig. 6.15. The now-space of a moving frame is tilted relative to the now-space of one at rest.

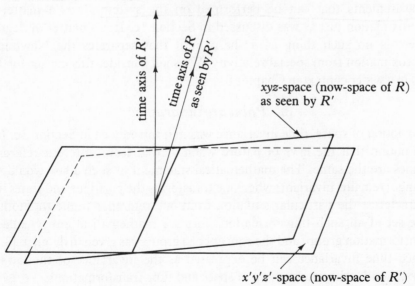

Therefore, *every event outside the light cone is simultaneous with the origin when viewed in a suitably moving frame*. This is the meaning of the space–time outside the light cone. It shows how space and time merge in an inseparable way: what is earlier for one frame is later for another; what is ordinary space for one is a mixture of space and time for another. *The separation of Minkowski space–time into space and time depends on the reference frame*. It is not possible to make an absolute separation of space and time. This is the ultimate message of relativistic space–time.

Finally, we turn to the relation between Minkowski space–time and Newtonian space–time. In Newtonian physics there is no limiting speed. The limit is infinity. We can therefore make our Minkowski space–time more and more Newtonian by making the speed of light larger and larger. This means that the angle of the light cone with the time axis must be made larger and larger until it reaches a right-angle. The cone becomes flatter and flatter, and in the limit of infinite light speed the cone flattens out completely and becomes the *xyz*-space. This is the Newtonian limit of space–time.

Since the Newtonian 'light cone' is flat and makes a right-angle with the time axis, all lines are now allowed as worldlines as long as they do not turn back in time. There is no longer a maximum speed. This flattening of the light cone is a very graphic way of seeing the relation between the special theory of relativity and Newtonian relativity. The transition is a gradual one: if the speed of light is 'for all intents and purposes' infinite, one has the Newtonian case. If all speeds in a physical system under study (for example, the solar system) are much less than the speed of light, no limiting speed is in sight and one has the Newtonian case. This has to do with the accuracy of the measurements that can be performed on the system. It is a matter of approximation just as was discussed in Section 1c. It is a matter of degree. There is no such thing as a sharp limit that separates the Newtonian approximation from special relativity. We shall consider this matter further and in a wider context in Chapter 8.

h. *Poincaré invariance*

The notion of space–time invariance was first introduced in Section 3c. It is the notion that the laws of nature when viewed from different reference frames are the same. The mathematical statement of such a law would not change (remains invariant) when one transforms the positions and times that characterize the particular situation from one reference frame to another. The set of all such transformations forms a mathematical entity called a transformation group, and the concept of a group was given in detail in [3.8]. Space–time invariance can be expressed as the invariance of the laws of physics under the whole group of space and time transformations.

Based on Galileo's law of inertia the particular class of reference frames called *inertial reference frames* was singled out (Section 5c). Only with respect to those frames do the laws of motion in Newtonian mechanics remain invariant. And the invariance group was seen to be the *Galilean group* of transformations. These involve relocations of the frame (translation transformations), reorientations of it (rotation transformations), and Galilean boost transformations. The latter relate two frames that move uniformly with respect to one another.

But in the present chapter we discover that those Galilean boost transformations describe the correct relation between inertial frames only when the speeds involved are small compared to the speed of light ($(v/c)^2 \ll 1$). Otherwise, the Lorentz boost transformations must replace the Galilean ones. And beyond that, the Principle of Relativity as stated by Einstein at the head of his special theory of relativity insists that all inertial frames should see the same laws not only in mechanics but also in the other branches of physics and in particular also in electrodynamics, Maxwell's famous theory of electricity and magnetism.

[margin handwriting: Galilean boost transformations only when velocity much less than c — otherwise Lorentz]

The characterization of inertial frames as related by the *Galilean* group of transformations of space and time is therefore only an approximation. One must generalize to a different group of transformations which contains the *Lorentz* transformations. It is intuitively very reasonable that this new group should also include the other transformations contained in the Galilean group: the translations and the rotations. Space continues to be homogeneous and isotropic by assumption (see Table 5.1 and Note [5.6]). When all these transformations are combined and also the time translations are added, the new group of transformations results. It is known as the *Poincaré group* in honor of the great French mathematician who first investigated it.

From a mathematical point of view one can characterize the special theory of relativity by the Poincaré group of transformations. One could say that the theory which has the symmetry that its laws are invariant under that group is the special theory of relativity. This is completely analogous to the case of Newtonian mechanics which is characterized by the Galilean group. But the (special) theory of relativity contains much more than mechanics as has already been pointed out above.

One can show that the Poincaré group of space and time transformations has a geometrical meaning which is exactly the geometry of Minkowski space–time that we have just seen in Section 6g. The power of mathematics permits one to establish such interrelations and to produce a conceptual coherence that would be very difficult to achieve in any other way. The resultant structure of the theory thus attains a certain internal consistency, simplicity, and beauty whose full appreciation would require the knowledge of the appropriate mathematical language. This may induce some readers to pursue this matter further.

Annotated reading list for Chapter 6

Bondi, H. 1964. *Relativity and Common Sense*. Garden City: Doubleday. A well-written, easy-to-read introduction into special relativity by one of the experts in the subject.

Durell, C. V. 1960. *Readable Relativity*. New York: Harper and Row. A thin and delightful little book for the beginner.

Einstein, A. 1961. *Relativity, the Special and the General Theory*. New York: Crown Publishers. A popularly understandable introduction by the master himself. His simple style compares favorably with most of his interpreters.

French, A. P. 1968. *Special Relativity*. New York: Norton Co. An excellent text but slightly more difficult than the above books.

Miller, A. I. 1981. *Albert Einstein's Special Theory of Relativity*. Reading: Addison-Wesley. A detailed account of the development of the subject written by a very competent physicist and historian of physics.

Pais, A. 1982. *Subtle is the Lord* . . . Oxford: Clarendon Press. This definitive biography of Einstein includes a complete scientific appreciation of his work. Despite its technical material it can be read very profitably by the non-expert by following the author's guidance.

Weisskopf, V. F. 1960. 'The visual appearance of rapidly moving objects.' *Physics Today*, **13**: Nov. pp.24–7. A very clear explanation of the impossibility of seeing a moving sphere Lorentz contracted written by an outstanding physicist who is also a fine popularizer.

7

Gravitation as geometry: general relativity

a. Newton's universal gravitation

The science of mechanics deals with the way in which forces determine the motion of bodies. It is an all-embracing science in the sense that it pays no attention to the nature or origin of the forces but only to their mathematical representation, their strength, their change with distance, etc. Whether these forces are of electric or gravitational origin, whether they are faithful descriptions or only crude approximations, or whether they refer to any force at all that occurs in nature, all this is quite irrelevant. Mechanics simply tells us what happens when a *given* force acts.

Other branches of physics deal with specific forces. They involve theories of forces due to magnets, due to friction, due to electric charges, due to gravitation, and so on. In all these theories one must employ mechanics in order to determine the motion that these specific forces cause. Newton's greatest work, the *Principia* (Section 5b), contains both his mechanics and his theory of forces due to gravitation. We have become acquainted with his mechanics and with its further development, Newtonian mechanics. Here, we need to review *Newton's gravitation theory*. It was hailed as one of the greatest achievements in the physical sciences. In fact, its fame grew with time as more and more evidence accumulated in its favor. Let us recall a few highlights.

Experimental and mathematical studies of the effect of gravitation on objects such as stones and cannon balls were initiated most actively by Galileo. He discovered the laws that govern the motion of free fall, that govern the motion of projectiles, that govern the motion down inclined planes, and so on. The motion of celestial objects, the sun, the moon, the planets, all of which he also studied, were quite a different matter.

The accumulated knowledge of millennia resulting from the human fascination with the celestial sky had culminated in Ptolemy's great work, the *Almagest*. It is the detailed and very complicated account of the motion of the sun, the moon, and the planets in a geocentric system. These motions were all reduced to what the ancient Greeks already knew as perfect motion: motion in circles. Then came the Copernican revolution which dethroned the earth and put the sun in the center of the system (1543). And then came the very extensive and very precise observations by the astronomer Tycho Brahe that enabled Johannes Kepler to discover his famous three laws of planetary motion (1609 and 1619) [7.1]. But in all this there was very little discussion of *forces* between the celestial bodies. What counted was the way they moved, *the geometry of their orbits*. In any case, there was clearly no conceivable relationship between the motion of celestial bodies and the motion of stones and cannon balls.

Newton's *Principia* changed all this (1687). He postulated one single and rather simple law of force,

$$F = G\frac{m_1 m_2}{d^2},$$

unification of forces on earth and in space

and he demonstrated the tremendous generality of it. The same constant G is used for *all* applications! That very same force is responsible not only for the motion of the heavenly bodies but also for the gravitational motion here on earth where it accounts for the motion of apples falling from trees [7.2] just as well as for the projectiles shot from the machineries of war. It is a truly *universal law*.

The meaning of the above formula is easy to understand: any two bodies with mass (m_1 and m_2 in the above formula) exert an attractive force on one another. This force is proportional to the product of the masses and inversely proportional to the square of the distance d between them (*inverse square law*). This distance must be measured from the center of gravity of one body to the center of gravity of the other [7.3].

The force law of gravitation combined with mechanics permits, by mathematical deduction, the great unification referred to earlier (Section 4b). It contained as special cases the laws found by Galileo as well as the planetary laws found by Kepler. But it accomplished a great deal more. Thus, for example, it also enabled an explanation of the tides as resulting from the attraction by the moon. It proved itself over and over again in the centuries since then, and it was in some instances confirmed in most spectacular ways. One of these was the discovery of a hitherto unknown planet.

The exact determination of the motion of the planets starting with Newton's law of gravitation was no easy task. The force of attraction exists not only between the sun and each planet but also between all the planets, each of which exerts an attraction on all the others. Nevertheless, the mathematical techniques were sharpened to permit enormous accuracy.

When the outermost planet then known, Uranus, showed a discrepancy between the calculated and the observed orbit, it raised a very serious problem. But two astronomers, John Couch Adams and Urbain Jean Joseph Leverrier, were able to carry the computations far enough to make a most startling prediction. They claimed that the discrepancy can be fully accounted for if there exists yet another planet, farther out than Uranus, whose attraction causes these discrepancies (1845). They also predicted the location of that planet, and it was indeed found in the following year close to where it was predicted to be (within one degree of arc). That planet became known as Neptune.

There is no theory in the physical sciences that is better established than Newton's theory of gravitation.

b. *Why Einstein searched for a new gravitation theory*

In 1922 Albert Einstein was awarded the Nobel prize for physics. But he had made a commitment to lecture in Japan just at the time when he would have had to attend the ceremonies in Stockholm. Typically, Einstein did not break his commitment but lectured in Kyoto on the subject: 'How I created the theory of relativity' [7.4]. In this lecture he presented two reasons that made him search for a new theory of gravitation.

The first reason is a fairly apparent one: his special theory of relativity is applicable to all forces known in nature except the gravitational force. From the mathematical point of view this is easy to see. Special relativity requires invariance of all fundamental laws under Lorentz transformations of distances and time intervals so that all inertial observers agree on the same laws. The above universal law of gravitation does not satisfy this requirement.

The second reason is a much deeper one. In order to construct an inertial reference frame one must be able to eliminate all forces so that a body can be verified to remain at rest or to move with constant velocity (the law of inertia). Gravitation cannot be eliminated like other forces can. For example, a body that is electrically charged will exert forces on other charged objects but one can neutralize that body's charge and thereby eliminate these forces. No such procedure is available for gravitation: one cannot neutralize the mass of a body, and one cannot, therefore, eliminate in this way the gravitational forces it exerts.

The clue to a solution of this problem came to Einstein when he realized that a falling person would not experience a gravitational force. His *fall eliminates gravitation*. If all the other kinds of forces are also eliminated it then follows that this falling person is an inertial observer! But for such an

[handwritten margin notes:]
① according to special relativity, all observers must agree on laws of nature — gravitation is the one they don't agree on

② can't eliminate gravity like other forces

(b/c fully
exposed to it)

falling
eliminates
gravity, but
then in
acceleration

observer (more precisely: for such a reference frame) the rest of the world passes by in an *accelerated* fashion. And two such inertial frames, one in North America and the other in Australia, say, would surely not move uniformly relative to one another. They would be in relative acceleration.

Such considerations do cast doubt on the whole framework of special relativity since that theory is based on inertial reference frames that can move *only uniformly* relative to one another. It restricts the validity of the physical laws to such inertial frames only. We suddenly realize that we have gotten rid of absolute *velocity* only to be trapped into a theory of *absolute acceleration*. Any frame that is not inertial is necessarily accelerated and the laws do not hold relative to such frames.

Is acceleration absolute? Newton certainly thought so. And he had a simple experiment to back it up. Take a pail of water and rotate it fast about its symmetry axis. The water surface will take on a concave shape which can be clearly distinguished from the plane surface of water at rest. Does this not prove that rotational acceleration can always be detected and is therefore absolute?

Fig. 7.1. Two pails of water on the same axis; one is rotating relative to the other.

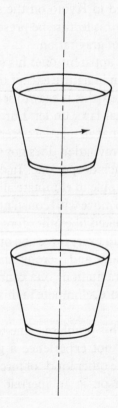

A strong opponent of absolute acceleration was the Austrian physicist Ernst Mach (1838–1916). He believed that only relative acceleration is meaningful. His argument can best be given by means of an extension of Newton's pail argument: suppose we have two such pails of water, one above the other attached to the same axis (Fig. 7.1). One of them is rotating relative to the other one. But if there were nothing in the universe except those two pails, how could one decide which one is at rest and which one is rotating? Symmetry would demand that both surfaces be the same.

Mach: relative acceleration (only compared to rest/uniform motion)

Such considerations convinced Mach that absolute acceleration is not a reasonable concept. But that leaves the question about what is responsible for the concave surface of a rotating pail of water as we do observe it? He gave a *dynamical* explanation for that. He argued that by and large all the matter in the universe, the distant fixed stars, the galaxies outside our own, all are in good approximation and on average at rest with respect to the earth. Therefore, when we rotate our pail of water we are doing so relative to all that distant matter, and the gravitational forces due to that matter cause the water surface to take on its concave shape.

Einstein was strongly influenced by Mach's ideas. He called the above conjecture *Mach's Principle*. However, to this day nobody has been able to provide a definitive theoretical or empirical proof of it. Many people though consider it very plausible. It certainly influenced Einstein's thinking; he felt that uniformly moving reference frames are much too restricted a class of frames and that the relativity principle of special relativity should be extended to reference frames that can be in arbitrary acceleration with respect to one another. This leads to the *Principle of General Relativity* which we shall consider in Section 7e.

c. *The equivalence principle*

We must now pursue this simple thought mentioned above: one can eliminate the action of any gravitational force by letting oneself be fully exposed to it, by letting oneself fall freely. In a sense, this was already known to Galileo when he discovered that all bodies fall at the same rate irrespective of their mass or composition. One can have a complete physics laboratory somewhere in the solar system falling freely under whatever gravitational forces might act on it. If that laboratory is not too large, all experiments in it can be done without any action of gravitation whatsoever. In fact, this is exactly what happens when astronauts work inside their space capsule.

Why must the laboratory not be too large? This follows from the following simple consideration. Suppose I sit in a closed room at a table and there are two apples on it one meter apart. If there were a deep elevator shaft directed straight toward the center of the earth and this room were falling down that

shaft, everything would be falling together. I would remain in the same
position relative to the table, and the two apples would remain on the table
(though weightless) at the same distance of one meter and falling at the same
rate. But the lines of fall all meet at the center of the earth so that all objects
would begin to move closer together. The two apples would not remain one
meter apart but would begin to approach one another. The rate of approach
would depend on the rate of convergence of the fall lines of the two apples. If
the apples are far apart this convergence is much more pronounced. Therefore, we want them so close together that we can neglect that convergence of
the fall lines during the short time that we are falling. This goes for all objects
in the laboratory, and therefore the room must be relatively small.

Like all space capsules, our room is also equipped with rockets for
acceleration. If our room were suddenly accelerated *upward* to stop the fall,
we would feel a sudden force exerted by the floor against our feet and our
feeling of weightlessness would disappear. But we would have the same
experience if for some mysterious reason the gravitational pull of the earth on
us suddenly increased. Could we tell the difference?

The *principle of equivalence* says that there is no way in which we can tell the
difference. If a *rocket engine* broke the fall of our room, the masses of all
objects in the room would provide the *inertia* to continue moving as before
(law of inertia), while the room, its floor, wall, and ceiling, would be slowed
down. If, instead, the *gravitational force* increased, the masses of all objects
would play their role in Newton's force law and would be attracted down
more than before; the masses will act as *gravitational masses* rather than as
inertial masses. The principle of equivalence thus asserts the equality of those
two kinds of mass.

The distinction between gravitational mass and inertial mass has the
following basis: in Newton's gravitational law of force as given above, the
masses determine the strength of the force. Without masses there are no
gravitational forces; hence the term 'gravitational' mass for them. On the
other hand, for all *other* forces Newton's force law $F = ma$ (force equals mass
times acceleration) tells us that mass here is a measure of inertia: the larger
the mass the smaller the acceleration a given force is able to produce. Hence
the term 'inertial' mass for it. Thus, one is really dealing with two quite
different concepts which have previously both been simply called 'mass'.
This important distinction lies at the root of the equivalence principle.

The equality of the two kinds of mass is a *necessary* condition for the action
of the rocket engine and the action of the gravitational force to yield equal
results and to be indistinguishable. The sufficiency for this to be true is an
even stronger assumption. Therefore one calls the equality of the inertial and
the gravitational masses the *weak principle of* equivalence, while the stronger

assertion that all phenomena inside the room will be exactly the same in the two situations is known as the *strong principle* of *equivalence*. they are identical

Precision measurements have succeeded in comparing the inertial and gravitational masses of various substances to high accuracy. They do indeed agree to the fantastic recent experimental precision of better than one part in 100 000 million! At the time Einstein developed his theory this equality of masses was measured 'only' to a precision of one part in 100 million. Such precision can of course not be claimed for the confirmation of the strong principle. But it too has been confirmed very well.

There is a remarkable consequence easily deducible from the principle of equivalence: the *bending of light rays* by gravity. This deduction can be based on a simple thought experiment.

Principle of equivalence leads to bending of light by gravity

Suppose we have a light source on the wall of our room. A narrow beam of light from there will cross the room and produce a bright spot on the opposite wall. But when our room is being pulled *up* by the force of our rocket engine, that bright spot will appear at a lower point of the wall than before. The displacement depends on the distance our room has moved while the light ray was in transit (Fig. 7.2). But at this point one can invoke the equivalence

Fig. 7.2. The equivalence principle leads to the conclusion that light rays are bent by gravitation.

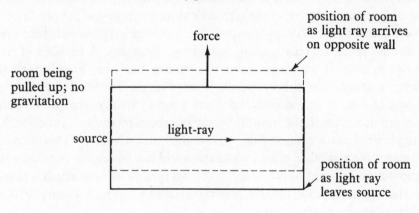

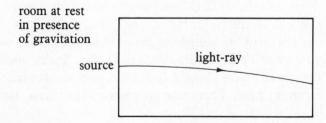

principle and assert that the same result could be obtained if a gravitational force would be acting *down* in the room. There is no way of telling the difference between gravitation (acting on the content of the room) pulling down and the rocket engine (acting on the walls, floor and ceiling) pulling up. The conclusion is immediate: a gravitational force (due to a big mass underneath our room) will deviate a light ray and pull it closer to that mass. In this way one concludes that light rays are bent by gravitation.

The bending of light rays is deduced by the above argument only in a qualitative way. After the general theory of relativity was developed, it provided quantitative predictions for this phenomenon. For the light ray from a distant star grazing the surface of the sun the bending is predicted to be 1.75 seconds of arc. This is a very small amount indeed but not beyond observability.

The first observation of the bending of light rays was a most dramatic event. It took place during a solar eclipse in 1919 and was organized by the Royal Astronomical Society of Britain. Only during an eclipse can a star whose light is just grazing the surface of the sun be seen. The photograph of that event was compared with a photograph of the same fixed star when the sun was far away. When the two photographs of the position of the star relative to the other stars in the vicinity were compared, a difference was indeed found. The presence of the sun near the path of the light ray produced an apparent displacement of the star indicating a bending of the ray.

Now at that time, 1919, the general theory of relativity was only four years old and accepted by only a small number of scientists. A number of rival theories existed of which we need to mention but two. First, there was Newton's gravitation theory which is so well confirmed. It predicts no bending of light at all. Second, there was a theory that combined Newton's law of gravitation with the result $E = mc^2$ of special relativity. Since light has energy it also has a mass according to that formula: $m = E/c^2$. This mass can be treated like any other gravitational mass and can be used in Newton's law of gravitation. The result is a bending of light rays by the sun which is *exactly half* the amount predicted by general relativity. Which theory will be confirmed by the observations?

The more conservative scientists made bets on either no deflection or half of the predicted 1.75. Newton's theory could not possibly be wrong. Tension mounted since it took several months after the eclipse to analyze the data of the eclipse observations that had been made at two different locations on earth. The British Astronomer Royal, Frank Watson Dyson, finally made the announcement. The results were about 2.0 and 1.6 with an estimated error of about ±0.3 seconds of arc. There was no question that these data excluded 0.87 (half the value) and confirmed the prediction of general relativity, 1.75.

The Times of London reported this as the 'most remarkable scientific event since the discovery of the planet Neptune' and announced the overthrow of Newton's ideas. In any case, it was this observation of the bending of light that established the general theory of relativity as an acceptable scientific theory despite its very unorthodox concepts (as we shall see), and made Einstein into the popular scientific idol for which he is generally known.

In very recent years, astronomers made discoveries that have a dramatic bearing on the gravitational bending of light. Using the best available telescopes (including radio telescopes) they discovered extremely distant objects with peculiar properties called *quasars* (for quasi-stellar objects). The light of some of these quasars passes on its way to us through a region containing very large masses (of the amount of a whole galaxy). As a result, their light is bent so that a *gravitational focussing* effect is obtained; those large masses act like a lens. One sees not only the quasar but also one or more additional images of it nearby in the sky. Calculations based on Einstein's gravitation theory confirm these observations quantitatively.

d. *Curved space–time*

The considerations that led to the principle of equivalence started with the observation that in a freely falling reference frame one cannot detect gravitation. This is a seemingly trivial observation. Yet, it implies that gravitation *is* detected whenever the reference frame is *not* freely falling. We experience gravitation because the rigidity of the earth prevents us from falling toward its center. While such a way of looking at gravitation is rather unusual, it does teach us an interesting lesson: it implies that *gravitation is a consequence of the choice of the reference frame*!

The principle of equivalence tells us, however, that this conclusion must be tempered: we may only *seem* to detect gravitation when in reality our laboratory is being accelerated by some non-gravitational force (such as a rocket engine) in the direction opposite to the pull of gravity. According to that principle, there is no way in which we can tell the difference from within our (small) laboratory.

When we want to check on the law of inertia by removing all the forces including gravitational forces, we must be freely falling in that gravitational field. But then we are *not* moving in a straight line. To be sure, the objects within our laboratory remain at rest relative to us or move with constant speed relative to us, and we *seem* to move in a straight line. All this is in agreement with the law of inertia. But from the outside it is quite evident that we are not moving in a straight line: the whole laboratory may be circling the earth. Thus, we conclude that the law of inertia holds relative to an *accelerated* reference frame. Gravitational acceleration is present because according to Newton the laboratory experiences a gravitational pull that makes it circle the earth.

This is a very confusing situation but matters are even worse than that. Recall that in the special theory of relativity light rays are assumed to move with constant velocity, that is with constant speed *and* in a fixed direction. In fact, a straight line can be *defined* by the way a light ray moves in free space (vacuum). But from the principle of equivalence we concluded that under a gravitational force light rays are *bent*: while they may keep at constant speed they do not keep at a constant direction. Gravitation seems to be incompatible with one of the fundamental principles on which the special theory was built: the constancy of the velocity of light.

These two examples present serious problems to the consistency of physical theory. They are exactly the kind of problems that Einstein struggled with during the 11-year period between his publications of the foundations of special relativity (1905) and of general relativity (1916). After an intensive and extremely difficult series of investigations he was finally led to the following solution: the paths followed by a freely falling laboratory or by a bent light ray actually *are* straight lines but 'relative to a curved space'.

This statement can only be understood after one defines what one means by 'straight' in such a way that it also applies to curved surfaces. 'Straight' is defined to mean 'along the shortest path'. Picture a sphere, the two-dimensional surface of a ball. Suppose that we were two-dimensional animals living on that sphere so that we would know only what is meant by forward and backward, right and left, but not what is meant by up or down. Then our notion of a 'straight' line would be a line that follows along a *greatest circle* on that sphere. (And that would not be 'straight' as seen by observers who live in three dimensions.) But the arc of a greatest circle is the shortest distance between two points on a sphere and in this sense is a straight line. In mathematics the line that one must follow in order to connect two points by the shortest path is called a *geodesic* line. In general, all shortest lines on *any* two-dimensional surface are called geodesics. Only when that surface is a plane will the geodesics be straight lines. In this way one generalizes the notion of 'straight' into the notion of geodesic when the surface is not flat as a plane.

The same generalization holds in higher dimensions and in particular in four-dimensional space–time. One can generalize 'flat' Minkowski space–time into a *curved space–time* called *Riemannian space–time* after the German

[handwritten margin note:] "straight" means "shortest path"

mathematician Georg Friedrich Bernhard Riemann. It corresponds to the generalization of a plane to a curved surface such as the surface of a relief map showing mountains and valleys. The shortest line in Riemannian space–time that connects two *events* is a geodesic line. This is an important generalization of the notion of 'shortest' since we are here dealing with space *and* time. But while this notion may present some conceptual difficulty, it is a mathematically well-defined notion that can also be expressed quantitatively for very general kinds of 'spaces'. And it is in this sense that in the presence of gravitation a freely falling body (small laboratory) follows a 'straight line', a geodesic, and a light ray also follows a 'straight line'.

Clearly, we have here a conceptual revolution. We are used to picturing the sun as a sphere placed in ordinary three-dimensional space and the light ray passing near the sun as curved because it follows a different line from the light rays that do not pass near the sun (which we call 'straight'). The Einsteinian view, however, pictures the sun as a 'depression' in an otherwise flat space–time so that the shortest paths (the geodesics followed by light rays) are straight only far away from that depression but curved nearby. And when one lives in that curved world (as do two-dimensional animals on two-dimensional surfaces) these shortest lines are straight lines.

Why don't they appear straight to *us*? Don't *we* live in this world? No, in an important sense we do not. Conceptually we do not live in this world because we separate time and ordinary three-dimensional space according to our own personal reference frames. If we were able to include time as a fourth dimension and were able to judge 'straightness' with that dimension included, we would indeed consider geodesics as straight lines. The problem is that when we think of space and time we do not take into account the other locations correctly: we ignore the fact that at other locations in space and time clock rates and meter sticks have sizes that *differ* from ours. We shall return to this important point shortly.

Einstein's solution of the above difficulties is therefore the following: the law of inertia is generalized to the statement that under gravitation, if there are no other forces acting, objects move along geodesics. Since absence of gravitation means that space–time is not curved, this statement reduces to the usual one about motion along straight lines in (flat) Minkowski space. Gravitation is thus represented as a curving of space–time: the sun is a depression in otherwise flat space–time.

It is not difficult to see that such a curving of space–time is in fact *demanded* when accelerated motion is combined with special relativity. Take the case of a rotating disk such as the platform of a merry-go-round. Contemplation of the seemingly simple mechanics of such a disk actually provides deep conceptual insight. It is a stepping stone for the transition from the world of special relativity to the world of general relativity, and it has in fact played that role historically.

We are assured that special relativity is valid relative to an inertial reference frame such as the non-rotating frame R outside the merry-go-round. Relative to R and using meter sticks at rest in R, the diameter D and the circumference C of the merry-go-round can easily be measured. One finds of course $C/D = \pi$ (which is numerically 3.141 59 ...) in accord with Euclidean geometry which is valid for R.

But we do not know which laws are valid relative to the rotating reference frame R'. Points on the rotating platform do not move with constant velocity in a straight line but have acceleration and move in a circle. But if the platform is large enough then a point on the rotating circumference C' moves almost in a straight line for a short while. We can therefore have a meter stick that moves exactly in a straight line (with constant velocity) move along with it so that it has for that instant exactly the same velocity as the point on C' (Fig. 7.3). In this way we can calibrate meter sticks at rest on C' with meter sticks moving uniformly in R. This permits us to transfer our standard of length from the inertial frame R to the non-inertial frame R'.

The uniformly moving meter stick will be contracted by a factor $1/\gamma$ according to special relativity. Therefore the standard at rest on C' which is a copy of it will also be contracted relative to one at rest in R. The observer in R' will measure along his circumference C' with a *shorter* meter stick than the

Fig. 7.3. A meter stick at rest in reference frame R' (rotating disk) is compared with a meter stick moving with constant velocity relative to an inertial reference frame R. At the instant shown both meter sticks have exactly the same velocity (in magnitude and direction).

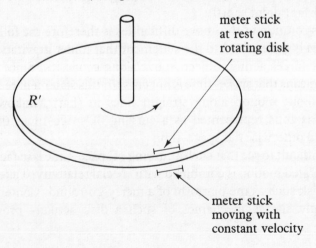

observer in R. Consequently, the observer in R' finds its circumference C' to be *larger* than C by a factor γ.

On the other hand, the radial distance on the platform will not be contracted because it is perpendicular to the direction of motion. A meter stick placed along the radius will measure the same as a meter stick at rest in R. The diameter D' of the platform as measured by the observer in R' will therefore be the same as the diameter D as measured by the observer in R, $D' = D$. The conclusion is that the ratio of circumference to diameter of that disk as measured in the rotating frame R' is $C'/D' = \gamma\, C/D = \gamma\pi$. This ratio is *larger* than plane Euclidean geometry permits.

It follows that in the reference frame R' which is uniformly rotating, the platform (which is at rest relative to R') appears not to be plane but curved. Euclidean geometry does not hold on it.

according to R',
R is curved
(doesn't follow
Euclidean geom.)

For non-Euclidean geometry the ratio of the circumference to the diameter of a circle can be smaller or larger than π. When we draw a circle on a sphere the ratio will be *less* than π; when we draw it on a saddle-shaped surface it will be *more* than π (Fig. 7.4).

Fig. 7.4. Circles are drawn on a plane, a sphere, and a saddle. The ratio of the circumference to the diameter C/D is equal to, less than, and greater than, respectively.

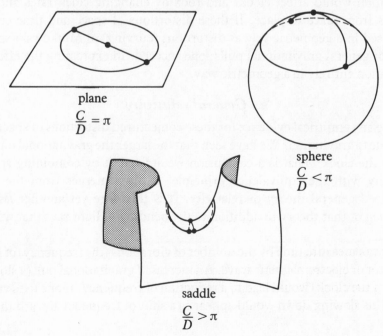

plane

$$\frac{C}{D} = \pi$$

sphere

$$\frac{C}{D} < \pi$$

saddle

$$\frac{C}{D} > \pi$$

In Newtonian mechanics the accelerations experienced by a rider on a merry-go-round are the fictitious centrifugal accelerations. The actual forces are pulling toward the axis of the merry-go-round forcing the horses to turn rather than to move in a straight line. Such forces are the product of *inertial masses* times acceleration. Therefore, by the principle of equivalence, they are (within a small region of space and time) indistinguishable from gravitational forces pulling outward. It is as if a large ring of mass were surrounding the merry-go-round and were to exert its gravitational pull. The centrifugal acceleration is in this view a (fictitious) gravitational acceleration. *Gravitation produces a distortion of space* which is needed to account for the larger circumference to diameter ratio we found above.

The argument just presented concerning length measurements relative to R and R' can easily be repeated for time measurements. One replaces the meter stick at rest on C' by a clock, and one compares that clock with one that is uniformly moving relative to R. One finds as before that the clock at rest in R' goes slower than the clock at rest in R. And the larger the platform the slower will be the clock on its circumference. Therefore also, *the larger the gravitational pull the slower the clock*.

Thus, the gravitational pull outward which a rotating observer experiences causes not only distortions of space but also distortions of time. And these distortions differ from place to place on the platform.

From this example it is not difficult to surmise that, quite generally, gravitation would affect clocks and rods by changing clock rates and rod lengths from place to place. If these distortions of space and time can be expressed in a geometric way as distortions (curving) of Minkowski space–time for general gravitational pulls, one succeeds in expressing the effect of gravitation entirely in a geometric way.

e. *General relativity*

Is there any empirical evidence for these conjectured distortions of space and time due to gravitation? We have seen that the larger the gravitational pull the slower the clock. That is a conclusion we arrived at by combining special relativity with the equivalence principle. It also emerges from the fully developed general theory of relativity. It is therefore yet another *testable prediction* of that theory in addition to the bending of light rays that we saw earlier.

One can measure time by the number of vibrations (the frequency) of some oscillator or electromagnetic wave. An increased gravitational pull (a slowing down of the clock) would lead to a reduction of frequency. If one uses visible light, this slowing down would appear as a shift of frequency toward the red color in the spectrum. One therefore speaks of a *gravitational red shift*. The larger the gravitational pull the larger the shift.

Over the years a variety of experiments and observations have been carried out to confirm such a gravitational red shift quantitatively. The oldest is the observation of light from dense stars on whose surface the gravitational acceleration is much larger than on the surface of the earth. The frequency of the light is indeed shifted toward the red by the predicted amount. But such observations have poor accuracy. Terrestrial observations involve a comparison of clocks at different altitudes. Their results also agree with the predictions and they have an experimental accuracy of about 1%. It is amusing to realize that the United States standard clock kept at the National Bureau of Standards in Boulder, Colorado, goes at a faster rate than the British standard clock kept at the Royal Greenwich Observatory since the former is at an altitude of over one mile more than the latter and therefore experiences a slightly smaller gravitational pull. The American clock is gaining about five millionths of a second per year over its British counterpart.

Summarizing our considerations, we can say that the general theory of relativity describes gravitation in a purely geometric way. It manifests itself as a curving of space–time. Every mass m changes the curvature of (Riemannian) space–time at the location of that mass, and this curvature gradually diminishes as one travels away from m. The solar system thus becomes a region of space–time which is curved in a complicated way. The two-dimensional analogy would be a sheet of rubber with heavy spheres of different sizes depressing it at the locations of the sun, the planets and all the moons. A freely falling space capsule among the planets in the solar system will follow a geodesic in this curved space–time. It may lead from regions of less gravitation (curvature) to regions of more gravitation. The apples on my table in this freely falling space capsule will move noticeably toward one another, making me aware of this change. *True gravitation* is present.

But we have also seen *apparent gravitation*. It is always present over small space–time regions where a change of reference frame to one that is freely falling will eliminate gravitation: if the space capsule falls for only a short period of time the principle of equivalence applies. No curvature is noticeable and space–time seems to be flat. Apparent gravitation can be 'transformed away' by simply changing the reference frame.

true gravitation can't be eliminated by changing reference frame

The geometrical meaning of this example of apparent gravitation is simply this: if one restricts oneself to a small enough region on a curved surface then it appears to be flat: the earth appears flat when one restricts oneself to the size of one's immediate neighborhood. Geometrically, a plane is a good approximation to a curved surface if one only looks near the point where the plane touches it. When this concept is applied to four-dimensional space–time one has the exact meaning of the equivalence principle.

But there are also situations involving apparent gravitation which are not limited to small space–time regions. Our rotating merry-go-round is an example. The apparent gravitation causing the centrifugal acceleration can be removed by changing one's reference frame to a non-rotating one, i.e. by getting off the merry-go-round. No centrifugal acceleration will be felt after that. On the other hand, true gravitation cannot be eliminated by a change of reference frame.

The geometrization of gravitation is perhaps easier to understand conceptually than it is to carry it out mathematically. What is involved can be sketched in the following way:

The presence of gravitation affects time intervals and spatial distances. The larger the gravitational pull the larger their change. What is a nice regular grid of straight lines in the absence of gravitation becomes a distorted grid of curved lines in its presence. In a plane one can first draw parallel lines in the x-direction and then, perpendicular to these, parallel lines in the y-direction. One obtains a grid resembling a street map of parallel streets in a city without hills. It is easy to locate any one point in the plane by means of such a *coordinate system*. One only needs to specify a starting point such as the center of town from where the houses are numbered. The geometry on this plane is Euclidean geometry.

When the plane becomes distorted by mountains and valleys the nice parallel coordinate system is lost. This can be seen best by picturing the plane as made of rubber and forced over a mountainous landscape. Lines that were straight are now curved, and lines that were parallel are no longer so. Neither do two lines that were perpendicular remain so. Such distortions are exactly what is produced when gravitation changes lengths and time intervals, except that we must then speak of space–time. Flat Minkowski space–time becomes curved Riemannian space–time.

There are an infinite number of different ways in which one can draw a coordinate system. Any arrangement of lines that permits one to find every point uniquely will do. Such systems are called Gaussian coordinate systems (after the German mathematician Carl Friedrich Gauss). Similarly, there are an infinite number of coordinate systems for a given curved surface. The choice of the coordinate system on that surface is clearly arbitrary. What matters are only the intrinsic properties of the surface, the shapes of the mountains and valleys. These coordinate systems play the role of arbitrary reference frames while the intrinsic features of the surface describe the realities of nature. The mathematical task of general relativity is to provide a description of nature so that the choice of the coordinate system has no physical significance. The laws of nature cannot depend on the choice of such a coordinate system or reference frame. These laws must be *invariant* under changes (transformations) from one such frame to another. This invariance property plays the same role in general relativity as Poincaré invariance plays in special relativity, or as Galilean invariance plays in Newtonian relativity. It is called *invariance under general coordinate transformations*.

The generality of this invariance includes reference frames that may be accelerated relative to one another. The special role that acceleration has played in special relativity (as being absolutely determinable) has thus been removed. There is now a 'democracy' established among the reference frames. *All* frames are equally good for the description of nature [7.5]. The goal of constructing a theory according to the general principle of relativity has thus been achieved.

The fundamental equations of the general theory of relativity state how the location of the masses in space–time determine the curvature (read: gravitational pull) everywhere in space–time. And knowing that, the motion of any object in such a curved space–time (read: under such a gravitational pull) is then also determined.

It follows that one can compute the motion of all the planets in the solar system by means of this new theory. The results are very nearly the same as those obtained by the old Newtonian gravitation theory. This is of course very satisfactory because that theory has been in such excellent agreement with observations. But there are some very small differences. Contrary to the belief first observationally motivated by Johannes Kepler and later deduced from Newton's law of gravitation, a single planet orbiting the sun does not exactly follow an elliptic orbit according to general relativity. Rather, it follows an elliptic orbit that precesses very slowly about the sun, known as the *precession of the perihelion* [7.6]. Can this small difference between the predictions by Einstein's and by Newton's gravitation theories be observed?

Such an effect has indeed been observed first in 1845 for the planet Mercury which is closest to the sun and for which it is expected to be larger than for any other planet. Newtonian gravitation theory had no explanation for it. Einstein's gravitation theory predicts 43.0 seconds of arc per century. The observed value agrees with this prediction within the error of the observations which is better than 1%. Later, the precession of the perihelion was also observed for Venus and Earth. These are also in complete agreement with the predictions of 8.63 and 3.8 seconds of arc per century respectively. This marks another triumph for general relativity.

Since these effects are so extremely small, one may well ask why it is that a theory so different from Newtonian gravitation theory gives almost identical predictions. One can prove mathematically that when the gravitational pull is relatively small (when space–time is only slightly curved), the fundamental equations of general relativity become, to very good approximation, just the fundamental equations of Newton's theory. As we shall discuss in much greater detail in Chapter 8, such a relationship between these two theories is not an accident. In fact, it is demanded by the consistency of physical theory in general.

f. *Gravitational radiation and black holes*

The three 'classical' predictions of the general theory of relativity, the gravitational red-shift, the bending of light, and the perihel motion of the planets, do not do justice to that theory for at least two reasons. Firstly because general relativity theory has tremendous fertility which is only barely touched upon by those three effects, and secondly because these effects may give the impression that all predictions of Einstein gravitation theory are very small corrections to Newtonian gravitation theory and are difficult to detect. In fact, if they were not small corrections, would not Newtonian gravitation theo.y have been found empirically incorrect a long time ago?

The answer to this last question is an emphatic 'no!' Newtonian gravitation theory has been empirically tested almost entirely on the solar system only. The rest of the universe has not been very well understood in terms of the laws of physics until relatively recently; nor has cosmology, the theory of the development of the universe, been understood quantitatively until the present century. The present era of cosmology did not start until 1929 when Edwin Powell Hubble discovered that our universe is expanding.

We want to present here only two more concepts entirely foreign to Newtonian gravitation theory: gravitational radiation and black holes. The latter will show how the presence of very strong gravitational fields leads to phenomena that can no longer be considered small corrections to Newtonian predictions. Both, gravitational radiation and black holes, are presently under very intensive scientific study so that our knowledge about them can be expected to increase considerably in the near future.

There exist a number of similarities between Einstein's general relativity and Maxwell's electrodynamics. As has already been pointed out, the universal law of gravitation (which emerges as a limit from general relativity) is of exactly the same form as Coulomb's law of attraction and repulsion of two electric charges at rest (which emerges from electrodynamics in the static case). Both are 'inverse square laws'. But there are also other similarities between general relativity and electrodynamics. One of these is the emission of radiation from moving sources.

In electrodynamics the sources are electric charges. When these are accelerated, electromagnetic radiation is emitted. Similarly, in Einstein's gravitation theory, the sources are the masses of bodies. These emit *gravitational radiation* whenever their motion perturbs the space–time curvature. Electromagnetic radiation consists of vibrations of electric and magnetic fields that propagate with the speed of light. Similarly, gravitational radiation consists of a vibrating gravitational field, i.e. small periodic changes in the space–time curvature. Indeed, gravitational radiation is a *vibration in the curvature of space–time that propagates with the speed of light*.

This picture makes gravitational radiation bizarre indeed. Is there empirical evidence for such a phenomenon? Unfortunately, even relatively very strong sources of gravitational radiation produce only barely detectable effects on earth. Direct observation is therefore still in its early stages of experimentation. But there exists indirect evidence of the emission of such radiation as has been observed recently.

A binary system consists of two stars that revolve about each other. If they lose energy by emitting gravitational radiation, their distance will gradually increase and so will their period, i.e. the time it takes for one of them to revolve about the other. These changes are very slow; in one year the period may change only by one part per billion. That is much too small a change to be observed. But in a few cases one of the two stars in such a binary system may be a *pulsar*, a very dense star that sends out electromagnetic radiation (for example X-rays) in pulses. In that case even such small changes can be measured.

A pulsar functions like a clock. But clocks are very sensitive to changes in gravitational pull because such changes cause changes in the clock rate, i.e. in the frequency of the pulses that are emitted. In this way the emission of gravitational radiation has indeed been detected indirectly: such emission causes an increase in the distance between the two stars so that the pulsar moves to a weaker gravitational field; that, in turn, increases the pulsar frequency which can be measured accurately. Observations of the pulsar frequency change agrees with the slowing down of the binary system as computed on the assumption that it is due to gravitational radiation loss. Another unbelievable phenomenon predicted by theory has thus been found to occur in nature.

The most dramatic predictions of Einstein's gravitation theory are *black holes*. This misleading name does not refer to holes but to regions of space in which the gravitational pull is extremely large, larger than a certain critical value. This critical value can be characterized as follows:

We already know that light rays are bent by a large mass such as the mass of the sun. Actually, that bending depends not only on the mass m but also on the distance d from the center of that mass at which the light ray passes.

The smaller the distance the larger is the bending. In fact, the bending depends on the ratio *m/d*. As stars burn out, as they deplete their energy so that they cannot radiate light any more, the gravitational pull of its matter collapses the star to a very small sphere of extremely high density. After our sun has exhausted its energy supply it will collapse to a sphere of only a few kilometers diameter. As a result the ratio *m/d* can increase tremendously: a collapsed star can bend light rays a great deal more strongly because the rays can pass by it at much closer range.

If a dead star has collapsed to such a small size that the light rays reaching its surface are *bent into the interior of the star* so that they cannot escape again, it is called a black hole (Fig. 7.5). Of course, if light rays get caught so does everything else (traveling at lower speed). <u>*Everything* that reaches a black hole remains in it forever</u>. Now a black hole does not send out radiation of its own; therefore, if it absorbs all radiation reaching it and reflects none, it is necessarily invisible. That region of space appears entirely black.

[margin note: black hole: collapsed star so dense light gets pulled in]

[margin note: black holes invisible b/c all radiation absorbed, none reflected]

Fig. 7.5. A black hole. A light source moving into a black hole sends out light rays. Their worldlines are not straight because of the curvature of space–time. Consequently, the light cones at different points have different orientations. The light rays sent out inside the black hole are curved slightly inward and cannot penetrate to the outside; the light cones all face inwards. No light can escape from a black hole. And since anything that moves slower than light must remain inside the light cone, nothing at all can escape (after Geroch, 1978).

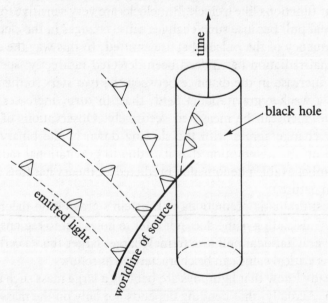

There is another way of looking at the creation of a black hole by collapse. On the surface of the earth it takes a certain minimum speed to escape the pull of gravity. This is why we need those enormous rockets to put men into space. That minimum speed is about 11 km/s. On the surface of the sun the escape speed would be much larger because of the much larger mass of the sun (its larger radius plays a lesser role). It would need to be about 40 times greater than on earth. As the sun collapses its radius decreases and the escape speed increases. When the radius of the sun reaches the small size of only 2.9 km (the critical size) that escape speed becomes the speed of light. At that point the sun has become a black hole because any further decrease of its radius ensures that not even light can escape from it. For the sake of comparison, the earth would have to shrink to a radius of less than one centimeter in order for the escape speed to become the speed of light.

How long does such a collapse take? If we were on the surface of the sun at the moment all its fuel had become exhausted and we were collapsing with the sun, the collapse would take only a few hours. But if we were on the earth, the collapse would take *forever*. Such is the difference in clock rates at the two positions.

What observational evidence do we have for the existence of such objects as black holes? How can we ever hope to see an invisible star? The situation is not as hopeless as it may seem. One can look for a binary system where only one of the two stars is visible but where that star behaves in such a manner that the presence of a second star in the vicinity can be inferred. If the behavior of the visible star and the associated radiation leads to the conclusion that the other star has the correct mass and size, a black hole can be inferred. Evidence along these lines has very recently become available in a few cases. While not entirely conclusive as yet, there seems to be a good chance that black holes are indeed being found in this way.

We have thus come to the end of a long journey. A complicated combination of empirical knowledge and pure thought has led from the gravitation theory of Newton to that of Einstein. The two theories are entirely different. One is based on a universal law of force, the other on an essentially geometrical reason for gravitation. That latter requires us to abandon many of our cherished notions of space and time, but it repays us in a deeper understanding of the world we live in. It accounts for all the gravitational phenomena we have known in the past and, at the same time, it provides us with a rich harvest of new and undreamed of predictions. To our amazement, these are indeed being discovered as improvements in our technology permit us to make increasingly precise observations.

Annotated reading list for Chapter 7

Bergmann, P. G. 1968. *The Riddle of Gravitation.* New York: Charles Scribner's Sons. An authoritative non-technical exposition by one of Einstein's former collaborators; it also includes a brief account of the special theory.

Cohen, I. B. 1960. *The Birth of a New Physics.* Garden City: Doubleday. A noted historian of science gives an excellent presentation of the developments in astronomy that led to Newton's gravitation theory.

Einstein, A. 1982. 'How I created the theory of relativity'. *Physics Today,* **35**: 45–7. This is a translation into English by Y. Ono of the 1922 lecture that Einstein gave (in German) in Kyoto, Japan. It is based on the notes taken by J. Ishiwara (in Japanese) who also gave a running translation to the Japanese audience.

Einstein, A. 1961. *Relativity, the Special and the General Theory. See*: Annotated reading list for Chapter 6.

Geroch, R. 1978. *General Relativity from A to B.* Chicago: The University of Chicago Press. A leading expert explains in popular terms the difficult geometrical aspects of the theory.

Greenstein, G. 1983. *Frozen Star: of Pulsars, Black Holes and the Fate of Stars.* New York: Freundlich Books. A fascinating narrative of the discovery and nature of these unbelievable celestial objects is told in non-technical language by a professional astronomer.

Pais, A. 1982. *'Subtle is the Lord . . .' See*: Annotated reading list for Chapter 6.

Sciama D. W. 1961. *The Unity of the Universe.* Garden City: Doubleday. A leading theorist explains in clear non-mathematical language and by means of many useful drawings how the observations of our universe lead to the basic principles responsible for its structure.

8

Revolutions without revolutions

a. *Established theories*

The term 'scientific theory' can mean many things. It can refer to theories that are currently believed to be true; it can refer to theories that were once thought to be true and that are now discarded; and it can refer to theories that some scientists believe to be true while others argue strongly against them. It is therefore necessary to be more specific.

To this end it is convenient to distinguish first of all between developing theories and mature theories, somewhat like one distinguishes between developing and mature persons. Well-articulated and well-confirmed theories that have survived serious and prolonged scrutiny by a number of experts in the field over many years of careful research are considered to be 'mature'. Before that, they are considered to be still in a phase of development. Of course, there is no precise instant after which a theory can be considered mature, just as there is no precise instant after which a person can be considered mature.

It may happen, however, that the large majority of scientists working in a particular specialty feel that a particular theory is very likely to be true even though it is far from well confirmed or is for some other reasons still in a developmental stage. In that case, we speak of an 'accepted' theory. Such a theory may not necessarily be a mature theory. Newton's corpuscular theory of light was accepted before its crucial tenets were put to decisive tests. The confirmatory evidence available at the time, together with Newton's authority, was considered sufficient for its general acceptance.

A mature theory can be expected to satisfy the criteria which were briefly discussed in Section 3b. Of course, depending on the situation, some of these may be better satisfied than others. It is, unfortunately, a matter of qualitative judgment rather than of quantitative criteria that leads to the acceptance of a theory and to its being considered mature.

But what happens when further advances in science lead to a new theory which supersedes the present one? Can this not also happen to mature theories?

The answer to this last question is an emphatic 'yes!'. Not only *can* this happen but it *must* happen as a matter of progress of science. It is exactly this point which is so often misunderstood by the non-scientist: every scientific theory, even the very best one, is *bound* to be superseded sooner or later. But this does not necessarily mean that the old theory is 'wrong' because the new theory is 'right', and how this can possibly be the case is precisely the subject matter of the present chapter.

While it may indeed happen (and *has* happened repeatedly in the history of science) that theories must be discarded in view of new evidence, there are many cases in which this is not the case and where the new theory provides for an *accumulation of knowledge* rather than for a replacement of it. If a theory must be discarded, the new theory takes its place. But when a theory survives, it does so within certain specified limits, its newly acquired validity limits. This means that the new theory is of importance only for a certain class of phenomena, namely those that lie outside those validity limits. Inside them, the old theory is 'good enough' which means that it is sufficiently accurate, that it is a good enough approximation.

The fact that the body of our scientific knowledge is steadily increasing, that there is *cumulativity in science*, is a clear indication that more theories are being superseded *without* being abandoned than are superseded *and* abandoned. But exactly how this can come about has been the topic of many a heated discussion among scientists and philosophers of science.

Perhaps the main stumbling block in this discussion centers around the notion that a surviving old theory is 'wrong' but that one can somehow get away with it while the new theory that supersedes it is 'correct'. This is a misconception and it must be clearly understood why this is so. The point is that no one scientific theory tells the whole story about nature and that all of the established theories are necessary.

Consider a mosaic floor made of very many small pieces of different color stones cemented together and depicting a beautiful scene. Such mosaics have survived from ancient Greek and Roman times and are of great value. Now such a mosaic is really nothing but stones and cement. But the knowledge of the location and color of each and every one of the little pieces of stone surely does not suffice to provide the artistic impression which the mosaic as a whole is meant to convey. Here, the whole is more than the sum of its parts. We are dealing with a *holistic situation*. While the mosaic can be 'reduced' to stones and cement, what the mosaic is all about cannot be 'deduced' from that.

Analogously, to take an example from the natural sciences, one can reduce the make-up of hereditary genes to atoms of hydrogen, carbon, oxygen, and nitrogen. But one cannot deduce from atomic physics all the content of genetics [8.1]. Nor does the requirement of vertical consistency (Section 3b) imply the requirement of such deducibility.

But let us now return to validity domains. We have already seen how a superseded theory which is not abandoned is given a restricted domain of validity. This domain is dictated by the superseding theory. An excellent example is the special theory of relativity. It supersedes classical Newtonian mechanics and restricts that theory to a domain in which the speeds of all participating objects are small compared to the speed of light. More precisely, the quantity $(v/c)^2$ is a measure of the error one makes when one applies the old theory (Newtonian mechanics). This quantity characterizes the validity limits: when it is so small compared to 1 that we are not concerned about the error we are making then the old theory is applicable. Otherwise, the new theory, special relativity, must be used. A validity domain does therefore not have sharp boundaries but is characterized by the size of the error one is willing to tolerate.

In this sense, no scientific theory is precise in the mathematical sense. *All theories are approximations.* But many of them are extremely good approximations. This important concept can be demonstrated easily by means of many well-known examples.

As these examples will show, each scientific theory is associated with a 'model' of nature, i.e. with a scenario made from components of our everyday experience which is supposed to resemble nature very closely. But these models necessarily only approximate reality. Therefore, they begin to fail when one goes beyond a certain 'domain of reality'. In the above example that domain is characterized by sufficiently low speeds relative to the speed of light. In all cases, the characterization involves a comparison of a quantity to another one, a ratio of two quantities.

Let us start our examples with models of nature which are much too simple to be called 'scientific theories' in today's sophisticated sense but which actually were considered to be such at the time they were first proposed:

'The earth is flat' was superseded by 'the earth is a sphere'. Even though we all believe today that the first assertion is wrong, we use it nevertheless as an excellent approximation: the surface of the ice of the frozen little pond is *plane* just like the surface of the water in my glass. On a spherical earth all bodies of still water are strictly parts of a spherical surface. But if the size of the object under consideration (the width of the pond) is small enough compared to the size of the earth (its radius), that spherical surface is approximated 'well enough' by a plane surface.

'The molecules of a gas are little balls that bounce against one another and against the walls of the container' is a model of nature which is superseded by a much more complicated model of molecules consisting of atoms bound together in various configurations. Nevertheless, the molecular theory of heat that is based on that model has provided us with great insight into the laws of gases. Of course, it also has its restricted domain of validity.

A much more sophisticated model is the one in which a sharp distinction is made between 'matter' and 'energy'. The validity domain of that model is characterized by all phenomena in which the energies involved are much less than the rest-energies of all particles involved so that no conversion can take place between rest-energy and other forms of energy. Such a conversion is possible according to $E = mc^2$ (Section 6f). But almost all phenomena encountered before about 50 years ago were of too low an energy.

In all these cases the superseded model was not 'wrong'. It was simply found to be valid only in a much smaller domain of validity than was originally thought. In the same way, Newtonian mechanics is not 'wrong' just because we now know about special relativity. It is just not a good enough approximation and is therefore restricted to a smaller domain than was previously thought. Theories which have thus been endowed with a validity domain and which were not discarded are called *established theories*. They are mature theories for which the validity limits have become known.

Established theories are preferable over merely mature theories because we know under exactly what conditions they are applicable. Thus we must be grateful to the superseding theory for having provided this information for us. Without that information we are never quite sure whether we are applying the theory to a situation for which it cannot give meaningful results. An example of that was the use of the Newtonian law of addition of speeds when the speed of light was involved. It led to contradictions until special relativity taught us that this law is not applicable for such speeds and that it must be replaced by the addition law for speeds of special relativity.

Finally, the superseding theory may have another important benefit for the superseded theory: it may eliminate from it various incorrect notions whose incorrectness one has not been aware of previously. In the case in point, special relativity eliminated the notion of the ether from the electrodynamics of Maxwell and showed that this notion is completely unnecessary and irrelevant. While Maxwell and his followers thought the ether to be an essential part of the model, Einstein showed this not to be the case. Today, nobody seems to miss it.

b. *Scientific revolutions*

The special theory of relativity was a scientific revolution. So was the general theory of relativity. So was quantum mechanics (we shall discuss it in Part C). The use of the term 'revolution' is chosen with justification. These were *conceptual revolutions* in the true sense of the word. For example, our old concepts of public space and time, even as conceived by scientists in the late nineteenth century (not to speak of the notions of absolute space and absolute time advocated by Newton) have been 'overthrown' as it were by the entirely different and strange concept of private space–time in special relativity. The two concepts of space and time, the old and the new, are quite incongruous and incompatible with one another. They are, according to a term used by the historian of science Thomas Kuhn, 'incommensurate', i.e. they cannot even be measured (compared) in the same way; they lack common ground.

Another example, even more in accord with the notion of a revolution of concepts, is the concept of *curvature of space–time* as set down by the general theory of relativity. That curvature is now held responsible for the motion of objects under gravity where we previously thought of *forces* doing that job. What can be more incongruous?

In view of these facts, it seems difficult to accept the notion that a scientific revolution can (and often does) leave the superseded theory intact except for the drawing of validity limits around it. How can *both* theories be 'right'? This issue has been studied and argued for many years and there is no general consensus on it [8.2].

Every scientific theory can represent nature only approximately. This is also true for the superseding theory. In due time, it, too, will be superseded by another one. Newtonian relativity was superseded by special relativity, and special relativity by general relativity. Quantum gravity will presumably some day supersede general relativity, and so on.

Each theory has its own model, its own concepts, and its own terminology. These are all chosen to optimize the approximation to nature which the theory represents. They form an internally consistent whole, possibly strengthened by a suitable mathematical structure. The crucial question, however, concerns the way in which these quite different theories are related. In particular, how is a superseding theory related to the theory that it supersedes?

The essential requirement here is that of *coherence* of these two theories. We called it *vertical coherence* in Section 3b. In probing vertical coherence the mathematical aspect of the two theories can be of very great help. Mathematics permits one to demonstrate how the equations of the more general superseding theory A lead to basic equations of the less general superseded theory B. This demonstration must be carried out by restricting the equations of A to the validity limits of B. It means that *in the approximation* in which B is valid, the mathematics of A becomes the mathematics of B. But the symbols of the mathematics of B have a quite different meaning from the symbols of the mathematics of A. In this way one can establish a mathematical link between A and B, thereby establishing a relation between the quite different concepts of the two theories.

A good example of this link occurs in the theory of heat. The macroscopic theory of heat (thermodynamics), theory B, involves the well-known concept of temperature. It is superseded by a microscopic theory of heat (statistical mechanics), theory A. In A there is no such concept as temperature. It deals with molecules and their motion. But the mathematics of the two theories is so related that one deduces from it a connection between the concept of temperature in B and the concept of average kinetic energy of the molecules in A. While these two concepts are indeed incommensurate, a very clear connection does nevertheless exist and can thus be exhibited by a comparison of the mathematics of the two theories.

But there are also situations where concepts of B theory emerge from concepts of the corresponding A theory when the latter is restricted to the validity domain of B. We have encountered one such example in Section 6f. When physical and chemical processes are restricted to low enough energies, the law of conservation of energy breaks up into two separate laws: the law of conservation of all mass energy and the law of conservation of the sum of all other forms of energy no matter how they convert into one another in the process; thus, the law of conservation of mass emerges in this approximation of low energies. Our insight into this matter in its historical development went of course in the opposite direction.

The mathematical relations between theories A and B are the basis for deducing theory B from theory A. But the concepts of theory B cannot all be deduced from the concepts of A, although there are some people who think so (*reductionists* in the strong sense). The emergence of new concepts in going from A to B is dramatized by the example of the mosaic floor given in Section a above.

The inability to deduce all of B from A is also one of the reasons why one does not use the superseding theory exclusively once it has become available. But perhaps a more important reason is a practical one: it is simply not feasible because of tremendous technical complexity without any benefit whatever. The superseding theory is always many times more difficult to apply to the same problems for which the old theory can be used with impunity. Thus, one can derive Kepler's laws of planetary motion from Einstein's gravitation theory. But any attempt to do it without first deducing Newton's gravitation theory from Einstein's theory would be a great deal more complicated.

B-type theories may contain holistic concepts. If they do, their concepts can certainly not be replaced by concepts of the corresponding A-type theories. That the whole can be more than the sum of its parts is also not an uncommon notion in the social sciences [8.3]. We shall encounter a different kind of holism in quantum mechanics.

These examples show what a scientific revolution does and what it does not do. If we concentrate on those cases where the old theory is not discarded then we find that such a revolution provides a completely new and different perspective for a wider class of natural phenomena than the old theory did. But at the same time, it strengthens the old theory by confirming its correctness within suitable limits and makes a mature theory into an established one. There is no contradiction between the new and the old, there are only different approximations appropriate to different levels of enquiry. The lack of contradictions is ensured by the vertical consistency of these theories. And beyond that, the superseded theory represents an important way of knowing which is not made superfluous by the superseding one.

Finally, it must be quite clear that the various approximate descriptions of nature provided by scientific theories all refer to the same real world. Our theories represent this world in different ways depending on the level on which we look at it, on the tools with which we investigate it, and on the domain of phenomena which we wish to include in our representation. All these established scientific theories are needed; they complement one another and together form a unified whole that is internally consistent. This consistency is assured by horizontal and vertical coherence between the various theories. It thus provides an ever increasing body of knowledge of what there is and a cumulative representation of nature in its great complexity.

Annotated reading list for Chapter 8

Kuhn, T. S. 1970. *The Structure of Scientific Revolutions*, 2nd edn. Chicago: University of Chicago Press. This very widely known and often quoted work has been extremely influential not only in the philosophy of science but also in various other areas. Its point of view has been attacked by some and supported by others since its first publication.

Polanyi, M. 1968. 'Life's irreducible structure'. *Science*, **160**: 1308. The author was a physical chemist as well as a social scientist. This article presents strong arguments in favor of a holistic view of life.

Rohrlich, F. and Hardin, L. 1983. 'Established theories'. *Philosophy of Science*, **50**: 603–17. A characterization of physical theories and their validity limits with a defense of scientific realism.

Part C

The quantum world

The belief in an external world independent of the perceiving subject is the basis of all natural science.

Albert Einstein

Natural science does not simply describe and explain nature; it is part of the interplay between nature and ourselves; it describes nature as exposed to our methods of questioning.

Werner Heisenberg

9

The limits of the classical world

The first quarter of our present century was a period of revolution in the foundations of physics. Startling new phenomena were discovered which could not be explained: the theories that had served science so well since Newton's time were suddenly found inadequate. Very strange ideas were advanced to account for these new experimental findings. First, the special theory of relativity and Einstein's gravitation theory shook some age-old beliefs while explaining some of them. But many more and even stranger phenomena needed to be understood. And that led scientists even farther beyond that familiar old world of physics that served so well for hundreds of years. The limits of *classical physics* had been reached and the era of *quantum physics* had begun. The present chapter sketches some of the highlights of this transition period.

a. *The classical world begins to fail*

The success of the physical sciences has been tremendous up to the present century. For three hundred years, since the time of Galileo, there occurred an unprecedented rise in science and technology. That period is now known as the 'classical' period in order to distinguish it from the new one, the 'quantum' period, which received its foundations in the mid-twenties of the present century. The classical fields of physics include the theory of motion of solid bodies when subject to given forces (mechanics), the motion of liquids and gases (fluid dynamics), the theory of light (optics), the theory of sound (acoustics), the theory of heat (thermodynamics and classical statistical mechanics [9.1]), the theory of electricity and magnetism including electro-magnetic radiation, and from early in the present century (see Part B) the theories of relativity, both the special and the general theory.

But already during the nineteenth century, observations were made for which the classical physical sciences had no explanation at all. Among the first of these was the discovery of *black lines* in the continuous spectrum of the sun.

As was first found by Newton, sunlight which seems to have no color, when passed through a glass prism, will fan out into light of different colors. This spectrum of colors is a continuous spectrum changing gradually from red through the colors of the rainbow to violet. But in the early 1800s it was discovered that under sufficiently high resolution (i.e. when that spectrum of colors is studied in sufficient detail) certain specific colors are absent and very narrow black lines appear in their place. The cause of this effect was not understood until 1859 when the German physicist Gustav Robert Kirchhoff found the explanation: as sunlight passes from the surface of the sun (the photosphere) through its outer layer (the chromosphere), some of the atoms and molecules of that layer absorb light of specific colors. The absence of these colors in the spectrum we see on earth appears as black lines. They are therefore called 'absorption lines'. Kirchhoff demonstrated by laboratory experiments that gases of different chemical elements absorb different colors which are specific to each element. This specificity is so perfect that one can use these absorption lines to identify and distinguish chemical elements. Conversely, when a chemical element is in the form of a gas of sufficiently high temperature, it will emit light of exactly those specific colors and only those. No explanation of this remarkable phenomenon was forthcoming from the scientific theories of the time: the otherwise so successful classical theories of physics failed to account for it.

A second problem that classical physics was unable to solve was the very fundamental problem of the stability of atoms. *What makes atoms stable?*

Even as late as the early part of this century, physicists knew very little about the structure of atoms. The original notion was of course that atoms are indivisible (that's the meaning of the Greek word 'atomos'). But the discovery of radioactivity had brought that ancient belief into question. Both negatively and positively charged particles had been observed to escape from radioactive substances. Now, the radioactive substances are electrically neutral so that each one of its atoms must also be electrically neutral. Therefore, radioactivity makes us believe that the neutral atoms must consist of a mixture of both kinds of electric charge, positive and negative. But just exactly how these charges are arranged inside an atom was at that time anybody's guess.

For example, the British physicist Sir Joseph John Thomson, whose achievements include the discovery of the electron, suggested that an atom might be a ball of positive electric charge in which the negatively charged electrons are embedded somewhat like raisins in a pudding.

Only the pioneering experiments by Ernest Rutherford and his collaborators in Manchester, England, in 1909–10 opened the way to a correct understanding of the structure of atoms. His fundamental and seminal experiments left little doubt that atoms have nearly all their mass concentrated in a center (the atomic nucleus) and that this center carries all the positive electric charge. The atomic nucleus is extremely small compared to the size of the whole atom, typically 10 000 times smaller in diameter. The negatively charged electrons move around in the space outside the atomic nucleus [9.2].

Once this nuclear structure of atoms (the nuclear atom) was established qualitatively, classical electromagnetic theory and classical mechanics were called upon to explain just how the electrons move around the nucleus of an atom such that the atom as a whole is a stable object. After all, opposite charges attract and like charges repel so that one has attractive forces between the atomic nucleus and the electrons and repulsive forces between the electrons. How can all these forces balance so that the atom as a whole is stable? It is exactly on this question that classical physics failed.

The electric force between a negative electron and the positive nucleus, the Coulomb force, is very similar to the gravitational force of Newton's universal law of gravitation: both are inverse square laws, the force decreasing with distance as the inverse square of the distance. The model of the nuclear atom therefore resembles very closely the Newtonian model of the solar system with electrons revolving about the nucleus in elliptic orbits, similar to planets revolving about the sun. But particles that are subject to a force are being accelerated, and accelerated charged particles radiate electromagnetic radiation. That is a necessary consequence of Maxwell's well- *according to* established classical electrodynamics (see Section 6a). And the emitted *classical physics,* radiation carries away energy. That makes the electrons revolve about the *electrons should* nucleus at closer and closer distance; eventually they fall into it. The atom is therefore not a stable configuration of particles. This situation is quite similar *fall into nucleus* to a satellite revolving about the earth. If it loses energy (for example because of friction with the earth's atmosphere) it eventually crashes to the ground.

That is the classical description of Rutherford's nuclear model of the atom: atoms cannot be stable according to classical physics.

Our last case exemplifying the failure of classical physics is the *photoelectric effect*. It was discovered in 1887 by Heinrich Hertz. He first noticed that ultraviolet light enhances a spark between two electric terminals. In a more refined experiment the terminals are encased in a glass tube from which the air is evacuated. When a battery is connected to these terminals, light that impinges on the negative terminal (the cathode) causes an electric current to flow in the circuit (see Fig. 9.1). One concludes that electrically charged particles must be moving inside the evacuated tube from one terminal to the other. After years of experimentation and after the discovery of the electron by the above mentioned J. J. Thomson (1897), it was eventually possible to establish that these particles are electrons which move from the negative to the positive terminal. The light impinging on the cathode somehow makes electrons emerge from it, and these then move through the evacuated space to the other terminal (the anode). They are thus closing the circuit and creating an electric current flow.

But what makes the photoelectric effect so puzzling to the classical physicist are its characteristic features. The most interesting of these are the following:

(1) The maximum energy of motion (kinetic energy) K of the liberated electrons increases in a straight line with the frequency f as shown in Fig. 9.2. The rate of increase is always the same no matter which metal is used for the cathode. All experimental curves of K versus f have *exactly the same slope*.

(2) The frequency f of the incident light (its number of vibrations per second) must be larger than a certain minimum frequency (threshold frequency) f_{th} no matter how high the intensity of that light. This threshold frequency, furthermore, depends very much on the kind of metal the cathode is made of. For f less than f_{th} no current flows.

Let us try to understand why these two features of the photoelectric effect are so puzzling to the classical physicist. In classical physics the energy of a wave (a light wave in our case) depends on its amplitude rather than on its frequency. Why then should an increase in frequency (without an increase in amplitude) transfer energy to the electrons? In fact, an experiment in which

Fig. 9.1. Vacuum tube for the photoelectric effect (schematic). The direction of the electric current is by definition the direction of motion of positive electric charges. The electrons emitted by the cathode are negatively charged.

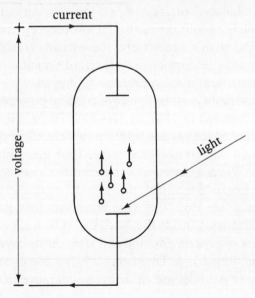

the amplitude of the wave is increased (keeping the frequency fixed) results in *no* increase of the electron energy K!

This accounts for the puzzle of (1) above. Concerning (2) one must first ask why there is a threshold. If it takes a certain amount of work (energy) to extract electrons from a metal one would expect that a certain minimum amount of radiation energy be necessary to make the current flow. That minimum is the threshold energy. But again, the threshold should then depend on the amplitude of the incident light rather than on its frequency.

The failure to explain the photoelectric effect, the failure to explain the sharp absorption lines in the solar spectrum, and the failure to account for an internal dynamics of atoms that permits them to be stable are three typical examples of the failure of classical physics which eventually led to the development of quantum physics. We shall encounter one more such failure in the following section.

failures of classical physics:
① absorption lines
② dynamics of atoms
③ photoelectric effect

It is of course clear that the beginnings of quantum physics did not wait for all these failures to have been brought to light. Rather, they developed simultaneously with the discovery of further failures. For example, as we shall see in the following section, the explanation of the photoelectric effect in terms of quantum physics was already available in 1905, while the experiments shown schematically in Fig. 9.2 were only done conclusively in 1916.

b. *The discovery of quantization*

It was clear that the inability of classical physics to account for various phenomena could not have meant that it had to be discarded. All these failures were related to the atomic world while classical physics continued to

Fig. 9.2. A classically unexplainable result. The maximum kinetic energy K of the electrons produced by light of frequency f increases linearly with f; but it is zero (no electrons emitted) if f is less than the threshold frequency.

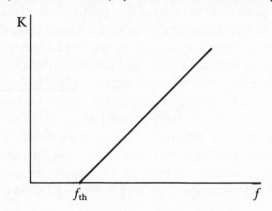

enjoy success and confirmation in the world of much larger scales than those of the atom. While it was surely not fully evident at the time, physics was confronted with applications of a theory (classical physics) outside its domain of validity, as was discussed in Chapter 8. At least in the atomic domain, classical physics had to be replaced by something quite different. Yet, no clue about the new theory existed. And when the first clue was in fact discovered, it was not even immediately recognized as such. The year was 1900.

The story is fascinating because the first clue did not at all come from an experiment on atoms. It had to do with something apparently quite remote from atomic physics: it arose in the comparison of the *theory of heat radiation* with experiment. Radiation is emitted by hot bodies and has been known since Hertz (see Section 6a) to be of an electromagnetic nature.

While all hot bodies radiate heat, the spectra of their radiation cannot easily be compared. The temperature of different parts of a hot body is not always the same and the heat radiation depends very strongly on the temperature as well as on the composition of the body. For this reason it is necessary to consider hot bodies that are *in thermal equilibrium* with their radiation. All parts of the body are then at the same temperature. Since in that case the radiation involves *all* wavelengths to a greater or lesser extent, the characteristic feature is the *radiation energy spectrum*. Such a spectrum exists for every given equilibrium temperature. It shows how the energy of the radiation is composed of the energies of radiation at all the different wave lengths. The spectrum is independent of the particular substance that acts as the source of the radiation and depends only on its temperature. A meaning-ful comparison of different experimental situations can then be made.

An excellent example of radiation that is in thermal equilibrium is the interior of a kiln such as is used for the firing of pottery. If the temperature in its interior is uniform the heat radiation emitted from the kiln will have an observed spectrum like the ones shown in Fig. 9.3 for 4000, 5000, and 6000 degrees. All curves show a spectrum shape that has one pronounced peak. For different temperatures the shape of the spectrum remains approximately the same, but for higher temperatures the curve is more peaked and is shifted so that its maximum occurs at higher energies and shorter wave lengths. These curves can be measured with considerable accuracy. The technical term for such radiation in thermal equilibrium is *black-body radiation* [9.3].

We see that the range of wave lengths that is visible (between about 4 and 7 of the indicated units) occupies only a small fraction of the wave lengths that are present in black-body radiation. Those extend from very short wave lengths (far in the ultraviolet) to very long ones (far in the infrared).

Now there was a serious problem in explaining the observed energy spectrum of heat radiation. For very long wave lengths the classical theory (worked out by Lord Rayleigh in 1900 and Albert Einstein in 1905) was in excellent agreement with experiments. This is indicated in Fig. 9.3 by the convergence of the curves marked 'classical theory' with the experimental curves (marked 'observed') for very large wave lengths λ. For short wave lengths, however, especially in the ultraviolet and in the visible range, experiments showed *qualitatively* different behavior: a maximum is reached followed by a fast decrease of energy with decreasing wave length. The classical theory shows a steady increase of energy as one goes to shorter and shorter wave lengths. In the limit of zero wave length the classical theory predicts the energy to be infinite. This classical prediction made no sense at all; it was a disaster for the theory and the problem was dubbed 'the ultraviolet catastrophe'.

incorrect prediction of black-body radiation

In 1900 the experimental curves in Fig. 9.3 were, however, correctly accounted for by a theory proposed by the German physicist Max Planck. He

Fig. 9.3. Heat radiation curves (black-body radiation). Each curve refers to a fixed temperature. It shows how the energy of heat radiation is distributed over the different wave lengths λ. The observed curves are shown as well as the curves predicted by classical theory. The temperature is given in units of degrees Kelvin, i.e. centigrades plus 273 (see Section 11g). The wave lengths are given in meters times 10^{-7}. The range of wave lengths for visible light is indicated. The agreement between theory and observations is very poor but improves for very long wave lengths. But even for 10^{-5} m (100 in the diagram), and at 10 000 degrees Kelvin, the theory is about 7 per cent too large.

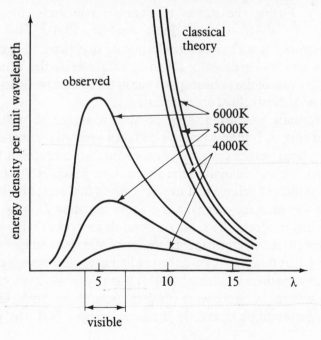

discovered a formula that can reproduce these curves exactly. The theoretical
arguments that Planck used to justify his formula are complex; they contained ingredients from the theory of heat as well as from electromagnetic
theory. The details are not important for us. What matters is that his
calculations had to deviate from the classical theory in two ways: he was
forced to assume *a smallest unit of radiation energy*, and he was forced to
assume that these smallest units are indistinguishable from one another.
While he was rather doubtful about the physical meaning of his assumptions
he realized that they were necessary 'to get the right answer'. In any case,
with those assumptions added to the classical theory he was able to 'derive'
his formula. And that formula is in excellent agreement with observations. It
is known today as *Planck's black-body radiation law*.

As a by-product of this mathematical derivation he found the strange and
totally unexpected result that for radiation of a given frequency these smallest
units of radiation energy have to be proportional to the frequency of the
radiation. The total energy is then a large multiple of that smallest unit since
the typically observed radiation contains enormous amounts of those smallest
units. If we denote the energy of these smallest units by E, and the frequency
by f, then

$$E = hf.$$

The letter h stands for a constant of proportionality; it was unknown and had
to be chosen to fit the experimental data. It is the only adjustable parameter
in his theory. Fitting the curves of Fig. 9.3 one finds (using today's
experiments) $h = 6.626 \times 10^{-34}$ Joule seconds [9.4]. This constant
became known as *Planck's constant*. Despite its questionable beginnings it
rose to fame and was eventually found to characterize the whole atomic
world; h plays a role of the same importance in the atomic world as the speed
of light c plays in the world of special relativity.

Planck's formula was astonishing because it agreed so well with the
experimental curves. But how could it be taken seriously if unfounded and
ad hoc assumptions were necessary to provide for its theoretical justification?
In fact, Planck himself thought of the above assumptions as a purely formal
device, and he did not believe that nature really is that way. Similarly, most
physicists at the time did not take the above equation $E = hf$ seriously,
though they had to accept the resultant formula as a remarkably good fit to
the data. After all, the electromagnetic theory of Maxwell and Hertz was well
established by that time and it described radiation as electromagnetic *waves*.
Uncounted observations confirmed that. A wave, after all, has a continuous
structure; how then could its energy consist of separate little pieces? And why
would these pieces all be of exactly the same energy? And why would the

energy of each piece be proportional to the frequency *f*? As we have already seen in the previous section, the energy of a wave in classical physics is known to depend on the amplitude, and it is independent of the frequency; the energy therefore cannot be proportional to the frequency. Thus there is a clear contradiction between classical electromagnetic theory and Planck's assumptions.

The little pieces of energy of radiation later became known as *light quanta*, Planck's assumption is called 'the quantization' of radiation energy, and his new theory 'the quantum theory of black-body radiation'. Of course, at the time this could hardly be called a theory.

Despite the cool reception by so many of Planck's great contemporaries, his quantum hypothesis received completely unexpected and vital support from a person who was quite unknown to the great scientists of the day; he was at that time a junior official in the patent office in Bern, Switzerland. Albert Einstein took Planck's result very seriously. He pictured radiation as actually consisting of these quanta, and he was extremely successful with this model. Not that he believed the quantum hypothesis to be a new fundamental insight; he considered it at that time to be of a somewhat heuristic nature. But in view of the experimental agreement he thought it to be a very good provisional assumption.

In 1905, earlier in the same year in which he published his special theory of relativity, Einstein gave his own derivation of Planck's formula. As a matter of consistency, he felt that the quantum model must also be used in the processes of emission and absorption of radiation by atoms. And this led him to the explanation of the photoelectric effect (Section 9a).

He reasoned that each light quantum incident on matter would either liberate exactly one electron from an atom or do nothing at all. In the first instance the quantum of electromagnetic energy disappears and its energy is used in two ways. Part of it is used to liberate an electron from the cathode by pulling it out of its bound state in the metal; the remaining part of its energy would be given to the electron as kinetic energy. In terms of symbols this reads

$$hf = B + K,$$

where *B* is the energy needed to free the electron from its binding, and *K* is the kinetic energy of the escaping electron. The equation simply expresses the law of conservation of energy.

This very simple idea explained all the features of the photoelectric effect, many of which were not even known at the time and were only found experimentally in later years. In fact, it was this work (rather than the theory of relativity) that earned Einstein the Nobel prize 17 years later.

Einstein's formula predicted two things:

(1) If one plots K versus f then it follows from the above formula that the slope of the resultant straight line is fully determined by Planck's constant h which is a universal constant. Therefore, no matter what the value of the constant B, all these plots have exactly the same slope.

(2) The threshold at which the photoeffect takes place, i.e. the energy which is just sufficient to liberate an electron from the cathode but which cannot provide it with kinetic energy, is the amount B. A quantum of that energy will lead to $K = 0$ and therefore have a frequency f_{th} which is related to B by the formula $B = hf_{th}$. Einstein's formula predicts a threshold frequency.

These predictions are of course exactly the features of the photoelectric effect which were indeed later found by experiment and which the classical theories of physics were unable to explain, as we saw in the preceding section (see also Fig. 9.2). Only when these features became experimentally established (about 1916) did the scientific community begin to take them seriously. These experiments also convinced Einstein of the *reality* of the electromagnetic energy quanta. He no longer considered them to be merely of a heuristic nature.

A new concept was added to this picture in 1923 when the American physicist Arthur Holly Compton carried out experiments in which he scattered X-rays on targets of various substances. He found that the scattered X-rays were of two kinds, some had the same frequency as the incident rays and some had appreciably lower frequencies. The latter emerged at various angles such that the lower the frequency the larger the angle by which they were scattered. The first kind of scattered radiation agreed with the prediction of the classical theory which treats X-rays as electromagnetic waves. The strange second kind, which had also been found by others was, however, interpreted by Compton in a most interesting way.

If Einstein's idea is correct, Compton reasoned, then this radiation must consist of individual quanta. These quanta would then *each separately* scatter from the individual electrons in the substance. The fundamental process pictured here is that of a collision between two particles: he treated each light quantum as a *particle* just like an electron. Simple calculations were then done by him (and independently by the physical chemist Peter Debye). They were based on this model of a two-particle collision and on the assumption that the electrons in the targets are essentially free, i.e. that their binding to the atoms is too weak to matter. The model required that the process obey not only the law of conservation of energy but also the law of conservation of momentum.

The result of these calculations showed complete agreement between theory and observations. The scattered quanta had indeed a lower energy exactly the way it was observed. Compton's experiments and calculations thus provided strong evidence for the *particle nature* of these quanta: they are *energy* quanta as well as *momentum* quanta.

But what about the part of the scattered radiation that agreed with the classical theory? For that part there must be a different model of the fundamental processes which *can* be correctly accounted for by classical physics. That model is the following. Since surely not all the electrons in the substance which scattered the X-rays can be free (or very weakly bound), one must distinguish scattering from free electrons and scattering from tightly bound ones as well as from atomic nuclei. The incident quanta scatter from the latter two kinds of particles as if they were very heavy. In that case, one concludes, the X-rays do not behave like a stream of particles (light quanta) but like classical electromagnetic radiation which is a wave motion.

Why electromagnetic radiation behaves like particles in one case and like waves in another was of course a complete mystery at the time. The two models are in clear contradiction with one another, and that problem could only be resolved several years later after the full quantum theory had been developed. We shall return to it at the end of Section 10f.

But what all this shows is that by the early 1920s it had become quite evident that the classical electromagnetic wave theory is not applicable in certain situations in the atomic domain and that radiation behaves in these cases not like a continuous wave as the classical theory claims but appears to be 'chopped up' into a stream of particles (quanta). Much more will have to be said about this 'schizophrenic behavior' of electromagnetic radiation appearing in some instances like a wave and in others like a stream of particles.

The conclusion of Compton's work, that in electromagnetic radiation not only the energy is quantized but also the momentum, can now be put into mathematical symbols: each quantum has energy $E = hf$ and momentum $p = hf/c$. But from special relativity follows for *any* particle of rest mass m, momentum p, and energy E that

$$E = \sqrt{[(pc)^2 + (mc^2)^2]}.$$

Consistency of these equations now leads to the deduction that $m = 0$: a light quantum is a particle that has *no rest mass*. No such particles had ever been seen before. They became known by the name *photons*, a term coined a few years later.

The discovery of the quantization of momentum was preceded by a proposal of an atomic model that required the quantization of yet another physical observable, the orbital angular momentum of an electron. Ten years before Compton's work established the photon, the Danish physicist Niels Bohr suggested a solution to the problem of the instability of the nuclear atom (see Section 9a). His solution, reminiscent of Einstein's solution of the problem of the constancy of the speed of light, was also achieved by fiat. Since the classical electrons orbiting around the nucleus of the atom lose energy by radiation and therefore fall into the nucleus, he postulated *stable* electron orbits in complete contradiction to classical electrodynamics. Bohr accomplished this stability by *postulating* that the orbital angular momentum of each electron be quantized, i.e. that it can occur only in integer multiples of a 'quantum' of angular momentum $h/(2\pi)$ where h is Planck's constant. If L denotes the orbital angular momentum of an atomic electron then, he demanded,

$$L = nh/(2\pi),$$

where n is an integer [9.5].

This postulate was remarkably successful. It accounted not only for the stability of atoms but also for the sharp spectral lines. Since it allows the angular momentum to change only by certain fixed amounts (quanta of angular momentum), the electrons can no longer orbit the nucleus at any arbitrary distance: only certain orbits are allowed, and these are at some distance from one another. An electron is therefore forced to 'jump' to get from one orbit to another, changing its energy in a *discontinuous* way. A higher orbit with more energy can thus be attained only when a photon of just the right energy is absorbed; a lower orbit with less energy can only be reached when a photon of just the right energy is emitted. Conservation of energy continues to hold. Thus, photons of very specific energy (i.e. frequency) are absorbed or emitted and the sharpness of the spectral lines of absorption and emission is explained. The mathematics of this theory gave a formula for the frequency of the spectral lines of hydrogen that agreed perfectly with the measured values (end of [9.5]).

Quantization thus resolved a series of problems that was beyond classical physics. It required an entirely new view. Quantities that were thought to vary continuously in classical physics such as energy, (linear) momentum, and angular momentum can, according to the view of quantum theory, change only discontinuously by certain minimum amounts. Electromagnetic radiation whether it be heat radiation, visible light, or X-rays behaves on the atomic scale as if it were a beam of particles (photons). Electrons in an atom are restricted by the quantization of angular momenta to move in discrete orbits; their jump from one orbit to another accounts for the sharp spectral lines. It is clear that the domain of applicability of classical physics is restricted and that quantum theory takes over beyond the limits of that domain.

But at this stage there was no new theory. There was classical physics modified by various *ad hoc* rules, the quantization rules. These were simply postulated 'to make things come out right'. That's not a theory. In the early1920s physics therefore found itself at the threshold of a major revolution. A completely new theory had to be found from which all these rules of quantization *follow* as a mathematical consequence. At the same time, that new theory had to be consistent with the classical theories of physics since those have served us so well outside the atomic domain and continue to do so. That new theory was *quantum mechanics*.

Annotated reading list for Chapter 9

Holton, G. and Brush, S. G. 1973. See the annotated reading list for Chapter 5.

Jammer, M. 1966. *The Conceptual Development of Quantum Mechanics*. New York: McGraw–Hill. One of the best resources on the history of quantum mechanics written by a highly respected historian of physics.

Pais, A. 1982. See the annotated reading list for Chapter 6. It contains a description of Einstein's view on quantum mechanics.

Pais, A. 1986. *Inward Bound. Of Matter and Forces in the Physical World*. Oxford: Clarendon Press. A history of particle physics from the end of the last century to the present. The first part (the history until the second world war) is less technical and is full of interesting details in the development of quantum mechanics.

10

Concepts of the quantum world

In the 1920s physicists discovered a theory which has been found to govern all phenomena of the quantum world which do not involve speeds close to the speed of light. This theory is called *quantum mechanics*. It plays the same role in the quantum world that Newtonian mechanics plays in the classical world.

The development of quantum mechanics led to the greatest conceptual revolution of our century and probably to the greatest that mankind had ever experienced. It most likely exceeded the great revolutions in our thinking brought about by the Copernican revolution, the Darwinian revolution, and the special as well as the general theory of relativity. Quantum mechanics forced us to reconsider our deepest convictions about the reality of nature. The profound philosophical questions it raised have even today, 60 years later, not been answered completely and in a satisfactory way. No wonder that the founders of the theory encountered great difficulties in accepting the interpretation of its mathematical structure. But the theory agreed so well with observations that they were forced to take that interpretation seriously despite its apparently weird character.

The foundations of the subject were laid down by a small group of people. Of the five most prominent ones three were about 40 years of age and the other two nearly a generation younger, in their early twenties. Two of the older ones, the Danish physicist Niels Bohr and the German Max Born, contributed more wisdom and interpretation than formalism or mathematical structure. That was done by the two younger ones, the Englishman Paul Adrien Maurice Dirac and the German Werner Heisenberg. The fifth one, the Austrian Erwin Schroedinger, played a curious role in that he contributed greatly and fundamentally to the formulation of the theory but turned his back on the interpretation which the others gave to it [10.1].

The background to the revolution is of course the long state of increasing surprise and wonder that had been going on for about a quarter century before. We sketched some of that in the preceding chapter. The various early attempts to construct a theory were generally recognized as fragmentary and unsatisfactory. A satisfactory and eventually triumphant theory finally did emerge. It came suddenly and within a relatively short period. Schroedinger developed his 'wave mechanics' and simultaneously and independently Heisenberg and Dirac constructed quite a different theory called 'matrix mechanics'. The two theories were apparently completely different. They differed even in the branches of mathematics that they used: wave mechanics used calculus while matrix mechanics used algebra. Both theories were, however, equally well confirmed by the experimental results. For some problems one of them was easier to apply and for other problems the other. But both gave the same answers for all the problems. There was simply no valid way to justify preference of one over the other.

The problem was soon resolved at least on a formal level: Schroedinger was able to show that the mathematics of either of the two theories can be 'translated' into the mathematics of the other. Rigorous mathematical foundations came in 1927 from the genius of the Hungarian-born mathematician John von Neumann, then an assistant to the great mathematician David Hilbert. He first found a very general mathematical basis for the theory and then demonstrated that this general basis permits various 'representations' that are mathematically fully equivalent. Two of these representations are wave mechanics and matrix mechanics. These two theories are therefore indeed completely equivalent. In fact, people discovered that it is advantageous to go back and forth between these two representations depending on the particular problem at hand. The gap between them had been bridged and the two theories merged. From that time on the general term *quantum mechanics* for both of them became standard usage in the scientific literature.

a. *Waves and particles*

Einstein's creative hypothesis that took Planck's light quantum seriously paid off handsomely: it led to the explanation of the photoelectric effect and to the discovery that these light quanta behave in every respect like particles. But these particles, the photons, give rise to a completely absurd picture. As we have already seen in Section 9b, electromagnetic radiation which had been well established by Maxwell's theory to be a wave phenomenon is now found to consist of particles [10.2].

The solution to this conundrum had to wait for quantum mechanics. The first spark came in 1924 in a doctoral dissertation by a French physicist, Louis de Broglie. According to him there exists a profound relationship between waves and particles so that waves can behave like particles and, *vice versa*, particles can behave like waves. He gave this idea a quantitative expression by proposing a mathematical relation between a particle property and a wave property. These two completely different concepts, particle and wave, are to be characterized by the momentum p for a particle and by the wave length λ for a wave. If one and the same object is to be both, a particle and a wave, then these two properties are related according to de Broglie's revolutionary hypothesis by

$$p = h/\lambda \text{ or } \lambda = h/p,$$

which are the same thing. Here h is Planck's constant.

As strange as this hypothesis appeared, it did agree with what was then already known, at least for the photon: its momentum p was indeed equal to h/λ because that is the same as hf/c ($\lambda f = c$). It is thus consistent with Compton's experiment as well as with special relativity (see Section 9b). But de Broglie's hypothesis also involved a prediction: not only are waves sometimes supposed to behave like particles (the photon being an example), but particles are sometimes supposed to behave like waves.

Here then was a good way of testing this unbelievable new idea. However, science does not usually proceed in a logical and systematic way. The experiments that indeed confirmed de Broglie's prediction were not at first carried out for the purpose of its confirmation. They were indeed started before that prediction became known [10.3]. In any case, the experiments by Clinton Joseph Davisson and collaborators (with C. H. Kunsman in 1921 and especially with L. H. Germer in 1927), which were done in Schenectady, New York, did provide a quantitative confirmation of the de Broglie relation. Another confirmation came from somewhat different experiments made also in 1927 by George Paget Thomson and collaborator (Andrew Reid) in Aberdeen, Scotland.

Fig. 10.1. The double-slit experiment. (*a*) Head-on view of the baffle; (*b*) cross-section through the baffle at *CC* and screen. A stream of incident balls is shown.

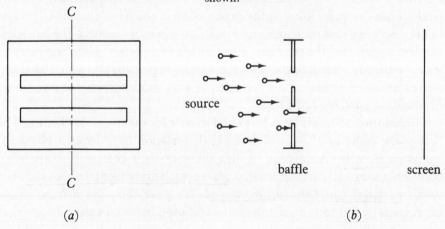

(a) (b)

When electrons were reflected from nickel crystals they did indeed behave just like waves. They showed interference maxima and minima just as radiation with X-rays would under similar conditions. In optics these interference effects are called *diffraction* and they had been well known for years [10.4]. But also electrons were now found to undergo diffraction. And, furthermore, the wave length that corresponds to that diffraction was found to agree exactly with de Broglie's formula.

Thus, there exists solid experimental evidence for both phenomena, that waves behave like particles and that particles behave like waves. But how can something be simultaneously a particle and a wave? Or are electrons particles on Mondays, Wednesdays, and Fridays, and waves on Tuesdays, Thursdays, and Saturdays?

Such sarcastic questions were indeed asked at the time when confused people tried to make sense of experiments that seemed to confirm absurd ideas. To find an answer to that problem, it is necessary to consider in detail exactly how a particle and a wave differ under the same experimental situation.

Let us therefore consider the following experimental set-up known as a *two-slit diffraction experiment*. The essentials are shown in Fig. 10.1. There is a baffle with two very long parallel slits. On one side, in front of the baffle, there is a source that can produce either particles or waves. A screen is placed on the other side far behind the baffle. The source can produce particles that arrive at the baffle with equal speeds and move nearly parallel to one another. The source can also produce a wave that will have a fixed wave length and a wave front that is plane and parallel to the baffle.

Consider particles first (Fig. 10.2). Each slit is assumed to be only about two or three times wider than the diameter of one of the particles. While all

Fig. 10.2. The double-slit experiment with classical particles (balls). The curves represent the number of ball impacts on the screen when both slits are open (*a*), and when only the upper or the lower slit is open (*b*).

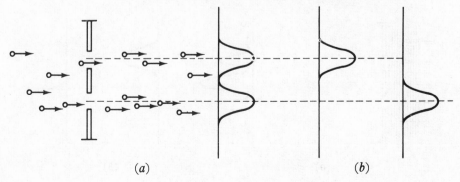

(*a*)   (*b*)

particles that do not hit the baffle will therefore come through the slits without changing direction, some of the particles will hit the edges of the slits and will be slightly deflected. The whole experiment is independent of the scale: we can imagine the particles either to be very small or to be as large as tennis balls as long as the slits are correspondingly of appropriate size.

What will be observed on the screen behind the baffle? If we count the number of impacts on each little square of the screen we find a distribution of impacts that has approximately the shape of a bell. It is drawn in the diagram as seen in cross-section. Fig. 10.2(*a*) shows the curve obtained when both slits are open; in Fig. 10.2(*b*) curves are given for the two cases when only one slit is open. The curve in (*a*) is just the sum of the two curves in (*b*).

Consider next the incidence of a wave, Fig. 10.3. In order to make this situation comparable to the previous one we shall assume that the slit openings are two or three times the size of the wave length. Again, the scale does not matter. The result will be the same for water waves as for light waves as long as the slit openings are of corresponding size. On the screen behind the baffle we record the intensity of the arriving light (or water). The screen will have very bright areas, dark areas, and areas that are only partly illuminated. Fig. 10.3 shows in part (*a*) the actual pattern of light intensity on the screen. It has its maximum in the center between the slits. On either side there are several smaller maxima. Looked at head-on we see a bright band in the center with several slightly less bright bands besides it, separated by dark bands. This is a typical diffraction pattern. It results from interference between the wave that comes through the upper slit and the wave that comes through the lower one. But when only one of the two slits is open, part (*b*), the intensity pattern shows no such interference, and it is similar to the one in the previous figure for particles (balls). As a consequence, the curve in

Fig. 10.3. The double-slit experiment with waves. The curves represent the light intensity on the screen when both slits are open (*a*) and when only the upper or the lower slit is open (*b*).

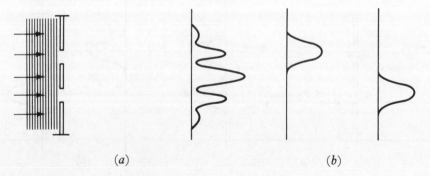

(*a*)   (*b*)

(*a*), when both slits are open, is *not* the sum of the curves in (*b*), when only one of the slits is open at a time.

What is the reason for the difference between the particle (ball) experiment and the wave experiment? The difference is clearly due to the *interference* which is present only for waves and not for particles. Let us therefore look very closely at this interference. It comes about when the crests of two waves meet and when they then form a crest that is double the height of each separate wave (constructive interference). In our case one wave comes from the upper, the other one from the lower slit. Or, a crest from the upper wave can meet with a trough from the lower one (or *vice versa*). In that case they cancel each other (destructive interference). These cases are depicted in Fig. 10.4. Which one of the two cases will occur, constructive or destructive interference, depends on the difference in the distances the two waves have to travel [10.5].

Interference can be described quite generally: the amplitude of the wave that results from interference is obtained by *the addition of the amplitudes* of two waves that meet and superpose one on the other. This is how the resultant amplitude of the two waves in Fig. 10.4 is constructed point by point. It is a very basic concept that underlies much of what follows.

The *intensity* of a wave is a measure of the amount of energy that the wave carries past us during one second. It is proportional to the *square of the*

Fig. 10.4. Constructive (*a*) and destructive (*b*) interference. The dash–dotted wave is reinforced by the dashed wave resulting in a wave of larger amplitude in case (*a*). The same dash–dotted wave is diminished by the dashed wave in (*b*) resulting in a wave of smaller amplitude (*b*). The effects are most pronounced when the two waves have equal maximum amplitude.

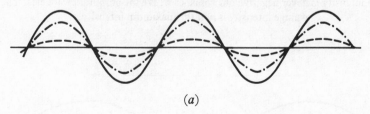

(*a*)

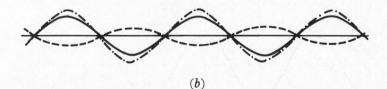

(*b*)

amplitude of the wave (Fig. 10.5). If a wave has three times the amplitude it will have nine times the intensity. In Fig. 10.3(b) the intensity of the light that comes through the upper slit is drawn. If A is its amplitude upon arrival at a particular point on the screen then the intensity at that point is A^2.

The intensity pattern of diffraction shown in Fig. 10.3(a) can now be obtained as follows. At each point on the screen one adds the amplitudes of the two waves that arrive there from the top and the bottom slits. One then computes the square of this sum. In symbols this can be stated very simply: if A is the amplitude of one wave and B is the amplitude of the other then the intensity I is given by

$$I = (A + B)^2.$$

We can finally make our comparison between the particle case and the wave case. When only one slit is open, the curve for the number of impacts of the particles (Fig. 10.2(b)) looks like the curve for the intensity of the wave (Fig. 10.3(b)). But when both slits are open we add the two 'intensity' curves for the particles (Fig. 10.2(a)), but we do not add the two intensity curves for the waves (Fig. 10.3(a)). Using the above symbols, in the case for particles we would compute $A^2 + B^2$; in the case for waves we would compute $(A + B)^2 = A^2 + B^2 + 2AB$. The two results are therefore different because the sum of the squares is not equal to the square of the sum. And this leads to very different curves in (a) of Figs. 10.2 and 10.3.

The understanding of the experiments that prove electrons behave like waves and those that prove electromagnetic waves behave like particles (photons) must be based on this knowledge.

Fig. 10.5. Amplitude A and intensity $I = A^2$. The wave length is also indicated. The intensity is never negative but vanishes where the amplitude vanishes. The average intensity is half the maximum intensity.

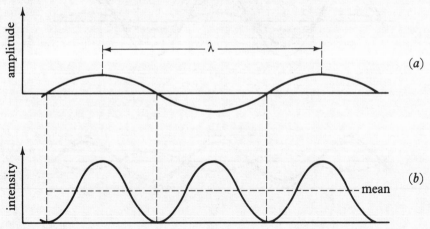

b. *Quantum particles*

The experimental confirmation of the de Broglie relation for electrons forces us to regard these particles as different from very tiny tennis balls but also different from waves. This is borne out by comparison of the two-slit experiment for these three: electrons, balls, and waves.

Let us start with electrons. Every electron that passes the two-slit baffle (Fig. 10.2) is recorded on the screen behind it by a flash of light from one point of the screen. (The screen is coated like a television screen for that purpose.) We must now keep track of the number of hits (flashes) at each point on the screen. We continue recording until we have a large enough number of hits so that we can draw an almost smooth curve showing their distribution.

The electrons behave like tiny balls rather than like waves when they each hit at just one point. A wave does not hit the screen at just one point; it produces a distribution of light all at once over the whole screen. But let us now look at the distribution that we have obtained by experiment from the stream of electrons when both slits are open. Does that look like the distribution of Fig. 10.2(*a*) for balls with both slits open? Here is the surprise: it does not look that way at all. Instead, it looks very much like the distribution of light intensity, Fig. 10.3(*a*), i.e. one that is produced by waves. The electrons behave like waves rather than particles as far as the *distribution* of hits is concerned.

Does that mean the electron behaves like a wave only in a statistical sense when one has a large number of them, as in the above stream of particles? This question can be answered by experiment. One can reduce the stream of electrons to a mere trickle so that there is only one electron at a time in the apparatus. The result is the same: one obtains the same interference curve (diffraction pattern) as with a heavy stream. Consequently, the effect does not depend on the rate at which the electrons hit the screen. The interference pattern is produced in a sense as if *each electron is able to interfere with itself*!

Despite the fact that each electron hits the screen at only one point (like a ball), we can associate a wave length with each of them as follows: we use the wave length of the wave that gives the same interference pattern as the electron. For a given width of the slits the wave length λ changes when one changes the momentum p of the incident electrons. The experiments then show that the momentum and the wave length are exactly related by the de Broglie relation $p \lambda = h$. This verifies de Broglie's conjecture and justifies the association of a wave length with each electron.

At this point it becomes desirable to introduce a denotational distinction: we shall call electrons or other particles that behave in the above way 'quantum particles' and distinguish them from 'ordinary' particles (balls) by calling the latter 'classical particles'. Both types of particles hit the screen in only one point; but the classical particles show a distribution of hits as in Fig. 10.2(*a*) (not an interference pattern) while quantum particles yield a curve as in Fig. 10.3(*a*), which *is* an interference pattern.

A quantum particle can thus be described by a 'wave' that interferes with itself. Of course, that would make the hit by an electron at a single point on the screen a mystery were it not for the fact that this 'wave' is not an ordinary wave. What kind of a wave is it? The essential feature that comes to our aid when we try to answer this difficult question is a fact to which we have so far paid no attention: we cannot predict *where* on the screen the electron will hit!

Actually, this is not surprising. For both, for electromagnetic waves and for classical particles we can predict exactly how the waves and the particles move through the slits and how they move to hit the screen. We have electromagnetic theory and classical mechanics for that purpose. These theories permit us to *predict* what will happen to each wave and to *each* classical particle. And this is why the patterns that result for those cases are not a cause of concern and puzzlement. But the interference pattern produced by quantum particles (e.g. electrons) is unexplained. We do not know how an electron can possibly move so as to produce an interference 'with itself'.

We are therefore no longer concerned with a *deterministic* dynamics as we have been for classical particles and for waves, but with a *probabilistic* one. In the latter we cannot determine the exact line along which an object moves. We cannot tell at which point the quantum particle will be after moving for a certain time. What we *can* determine is a *distribution of probability* which tells us the chance that the quantum particle will hit a given point. Let us therefore look at the outcome of the two-slit experiment from that point of view.

The result of throwing tennis balls (classical particles) through a single slit is given by Fig. 10.2(*b*) which is now interpreted as the probability distribution telling us where the ball is more likely to hit the screen and where less likely *without being able to tell us where any one ball is going to hit* (non-deterministic information). Similarly, the curve of Fig. 10.2(*a*) can be understood as a probability distribution for the case when both slits are open.

Turning now to quantum particles when only one slit is open, Fig. 10.3(*b*) can be interpreted as a plot of the probability that tells us for each point on the screen how likely it will be hit. Since it looks just like the one for classical particles, Fig. 10.2(*b*), one might think that it can also be obtained by a deterministic dynamics and that there is in fact no difference between quantum particles and classical particles. But proceeding to Fig. 10.3(*a*) we find for the two-slit case an interference pattern which can no longer be explained by the classical deterministic dynamics. A probabilistic interpretation is, however, still possible. Such an interpretation asserts, for example, that the *maximum* of the curve in Fig. 10.3(*a*), the highest probability of hits when both slits are open, occurs exactly in the center of the screen. For classical particles, in contrast, there is a *minimum probability* in that very place (Fig. 10.2(*a*)).

It follows that we must have a theory which permits us to *compute* the probability distribution from the knowledge of the velocity (or momentum) of the incident quantum particles and of the size of the two slits in the baffle. That is exactly what quantum mechanics is able to do. It provides a *probabilistic dynamics*. This means it provides equations not for the computation of the *position* of each particle as it changes in time (as would be the case for classical particles) but rather for the computation of the *probability* of being hit for each point on the screen.

But we know that an interference pattern can be explained only by the addition of the amplitudes of two 'waves', as we have seen earlier. We must therefore be able to compute an *amplitude* for the quantum particle which in some way resembles the amplitude of a wave. The essential difference is that the square of this amplitude [10.6] does not give the *intensity* distribution (as is the case for a light wave, for example) but it gives a *probability* distribution. By analogy, that amplitude is therefore called *probability amplitude*. It was Max Born who first suggested that quantum mechanics must be interpreted as a probabilistic dynamics.

The de Broglie wave of a quantum particle therefore has a probability amplitude in the sense that it is related to probability in the same way as the amplitude of an ordinary wave is related to its intensity. The wave length of this 'wave' is related to the momentum of the particle by the de Broglie relation. In the Schroedinger description of quantum mechanics this probability amplitude is called 'the wave function'. It is denoted by the Greek letter ψ.

The question 'how can a particle be a wave?' can now be answered in the spirit of quantum mechanics. A quantum particle is a wave in the sense that its whereabouts is described by a probability amplitude that is like a wave and that is capable of interference. Its square gives the observed probability distribution.

The classical physicist was confronted with the puzzle 'how can an electron be both a particle and a wave?' With this question he had in mind the classical notions of particle and wave. The quantum mechanical answer is: 'The electron is neither; it is neither a classical particle nor a classical wave. It is a quantum particle and as such it has some properties that resemble certain classical particle properties and others that resemble those of a classical wave.'

[handwritten margin note: electron neither wave nor particle: has props resembling both]

Sometimes people use the term 'dualism' referring to electrons as both particles and waves. This is a very misleading term since neither classical concept applies. The whole difficulty arose in the first place because one tried to use these concepts where they do not apply. There is no reason other than prejudice to expect the quantum world to be expressible in classical terms. Since that world is admittedly strange to us, being very far removed from our experience, it should come as no surprise that many of the problems we have in comprehending it are due to <u>our lack</u> of proper words for its <u>new and unfamiliar concepts</u> and for its peculiar nature. The analogies we can draw to the world we are familiar with are in general rather poor. This difficulty also existed in the world of relativity but to a much lesser degree. There we could model curved space–time by means of spaces in lower dimensions to help us hone our intuition. No such devices are available here. This is why the quantum world offers a very special challenge.

c. *Indeterminacy*

The world of the classical physical sciences is ruled by determinacy. The fundamental equations permit one to determine the future from the past in every detail from the knowledge of the initial conditions of the physical system. This is true for the mechanical aspects of things as well as for the electromagnetic ones. Both worlds, the one of Newton and the one of Maxwell, agree on this.

Physical determinism by no means originated with Newton. It goes back to the early atomists and the Epicureans. Those philosophers even argued that if all is made of atoms then their attributes and motion must also determine the behaviour of human beings. This started the old and very long debate on free will. Later, determinists argued that the world runs like a machine created by a higher being, and all that happens has been pre-determined at the beginning. This view received considerable support by Newtonian mechanics.

We shall call this view '<u>naive determinism</u>'. Apart from philosophical arguments against it, later scientific developments have also proven it to be untenable, even in classical mechanics, without at least a considerable amount of qualification. Let us sketch this matter very briefly.

Already a century ago, two great mathematicians, A. M. Liapunov and Henri Poincaré, pointed out how the equations of classical mechanics do *not* always determine the future motion of an object uniquely from given initial conditions. Examples of such situations are in fact well known to everyone. When a weight suspended by a rigid rod forms a pendulum that can oscillate back and forth, something unpredictable can happen. When that <u>pendulum</u> swings out with enough speed to move up and reach the vertical so that it is for a moment upside down, one cannot predict whether it is going to move back down the way it came up or whether it is going to continue and complete the circle. The pendulum's upside down position is a point of instability of the motion at which classical determinism fails.

Worse than that, in many instances the solutions of the classical equations of motion may change drastically when the initial conditions are modified only very slightly. Practical preparation of a physical system does not allow fixing these initial conditions (say the initial position and velocity of a rocket) to mathematical precision. Therefore, when the motion of the system is very sensitive to these initial conditions the distant future of such a classical system becomes completely indeterminate. In recent years classical physical systems whose future behavior is unpredictable and can even become *chaotic* (moving in seemingly random fashion), have been found to play an increasingly important role. The naive determinism that emerged from earlier Newtonian mechanics and that has been the basis of so much philosophical discussion is therefore not tenable any longer. It certainly does not hold in quantum mechanics. But it does not even hold in classical physical sciences in very many actual situations.

We have already seen how determinism in quantum mechanics is called into question. Following a conceptual rather than a historical line of argument the two-slit thought experiment has proved extremely valuable as an introduction to the problems of the quantum world and to the thinking it precipitates. We were first forced into the distinction between classical and quantum particles. Then we had to account for the interference pattern that a stream of quantum particles exhibits no matter how tenuous that stream. And that led us into a probabilistic description of the motion of these quantum particles.

One of the most interesting and surprising aspects of quantum mechanics is, however, the following claim: *the probabilistic description is also the fundamental description; there is no deeper level*.

This seems very difficult to accept. After all, classical mechanics also has a probabilistic description. For instance, one describes a gas in statistical terms. Since it consists of a very large number of molecules that move seemingly in a random way, colliding with one another as well as with the wall of the container, one treats them statistically. More generally, this approach, called *classical statistical mechanics* (see [9.1]), is applicable to *ensembles* of particles, i.e. to large collections of them such as a stream of electrons or a gas. It investigates the *statistical distribution* of the velocity, energy, and other properties of the particles that make up the ensemble. Given such a distribution of velocities (i.e. a curve that tells how many particles have which velocity) the average velocity can then be computed. This is not dissimilar to

[handwritten margin note: ☆ can only predict probabilities]

the way in which the average population density is computed from the distribution of the population throughout the country (high in and around cities, low in rural areas). And since there are such an extremely large number of particles involved, for example 10^{20} molecules in only one cubic centimeter of a gas, one works better in this statistical way. Following each molecule in its collisions with the others would be far beyond human capacity. *But* this theory is still classical, and it is therefore deterministic because *in principle* one *could* follow each molecule along its way. One just lacks all the detailed information necessary to do so. If all the initial data were available one *could* predict exactly how it moves through the gas as it collides repeatedly with the other molecules or with the walls of the container.

No such detailed mechanics is available in the quantum world. Quantum mechanics gives us the probability amplitude that was mentioned earlier, from which a probability can be computed. For example, one can compute the probability that an electron would have a certain velocity. But quantum mechanics does not give the velocity of a specific particle. And the claim is that no such detailed information even *exists*. The question itself, 'what is the velocity of a specific particle in an ensemble of particles?' is in most cases considered to be meaningless [10.7].

Needless to say, this state of affairs was not considered to be satisfactory by everyone. Attempts were made to correct it by constructing a detailed mechanics. To this end it was necessary to introduce attributes of quantum particles that are not observed. These were called *hidden variables* and the efforts in this direction became known as hidden variable theories. Although the effort has been going on for years those theories have either run afoul of experimental evidence or were considered too contrived to be generally acceptable. No empirical evidence exists in favor of as detailed a mechanism as they advocate. They are proposed entirely on the basis of philosophical arguments. Today there are only very few scientists who continue to support such theories [10.8]. We shall return to this issue in Section 11e.

But there is yet another contention of quantum mechanics. In addition to the absence of a detailed mechanics for an ensemble of quantum particles, quantum mechanics makes one more claim. It claims that its probabilistic mechanics describes *single quantum particles* such as single electrons.

There was, understandably, also resistance to this claim. Some physicists have argued that such a claim is unnecessary, and they have tried to give quantum mechanics a *statistical interpretation*. This means that they have tried to interpret the probability amplitude as a quantity that refers only to an ensemble and not to a single particle. This view is also not generally accepted [10.9]. After all, experiments show that each quantum particle can interfere with itself (as we have seen), so that it is at least reasonable to apply the

[handwritten margin note:] not just group described by probability, but each electron (after all each can interfere with self in two-slit)

probabilistic predictions to each individual particle. The probability *distribution* can of course be tested only by means of a large number of repetitions of the same experimental set-up.

It must be emphasized that, despite the indeterminacy of the behavior of a specific quantum particle, *precise predictions* can be made. In fact, the theory would hardly be taken seriously were it not for its ability to predict. Much of the theory's credibility derives from the confirmation of predictions. While not everything can be predicted, those things that can be are predicted in one of two ways: exactly (as in classical physics) or in a probabilistic way. Predictions of the latter type include the interference pattern in the double-slit experiment. It has been fully confirmed as accurately as it can be measured. Non-probabilistic predictions include the extremely precise values of the photon energies emitted from atoms; these have also been fully confirmed.

Finally, lest there be a complete misunderstanding of the implications of indeterminacy one must emphasize the existence of *causality* in quantum mechanics. Indeterminacy in no way precludes causality: not *all* questions have probabilistic answers. For example, in the photoelectric effect the emission of an electron is caused by the absorption of a photon. The fact that the location in the metal where this takes place is indeterminate and not predictable does not affect this causal relationship. The photoelectric effect is also controlled by conservation of energy. This conservation law, like others in classical mechanics, holds as strictly in the quantum world as in the classical one. Similarly, the transition of an electron in an atom from a higher energy state to a lower one is the cause of the emission of a photon of corresponding energy. The fact that the exact time of the occurrence of such an event cannot be predicted in no way detracts from this cause–effect relationship.

[*handwritten margin note:* cause-effect maintained]

In quantum mechanics indeterminacy, predictability, and causality in this sense exist peacefully side by side.

d. *Uncertainty*

Indeterminacy is not a matter of our inability to do better. Quantum mechanics claims it to be intrinsic to the nature of the quantum world. For example, in many cases the *location* of a quantum particle is given by a probability distribution. That a particle may not have a definite position is a fundamentally new notion. It can be substantiated by means of our double-slit thought experiment (Figs. 10.2 and 10.3) as follows:

[*handwritten margin note:* indeterminacy is intrinsic to world]

Since the electron hits the screen in just one point it is reasonable to assume that it has been localized all along and that it has been traveling in a line (not necessarily straight) as classical particles do. In that case it must have come through *one* of the two slits. In order to check on this, <u>detectors</u> which will localize the electron can be placed at the two slits, i.e. will <u>tell us through which slit each electron</u> traveled. Then one can separate the part of the interference pattern which is produced by the electrons that come through the upper slit from the part which is produced by the ones that come through the lower slit. As before, one must wait until a sufficiently large number of electrons has passed so that there are enough statistics to draw the pattern of hits on the screen and thereby obtain the shape of the probability distribution. But when all this is done the result is completely unexpected: when one observes through which slit the electron came, <u>the interference pattern</u> disappears! Instead of the quantum mechanical pattern Fig. 10.3(*a*) one finds the classical pattern Fig. 10.2(*a*). The attempt to localize the electron along the baffle destroys the interference.

Here is an example of how a change in the experimental set-up (the addition of detectors near the slits) can destroy the previously observed state of the system and can replace it by a completely different one. *The measurement of one property of the system* (its position) *destroys another property of the system* (the ability to interfere as evidenced in the interference pattern). One <u>cannot see both properties at the same time</u>.

[handwritten left margin: measuring electron's position destroys its ability to interfere.]

It is reasonable to suspect that the act of measurement of their positions has disturbed the electrons so much that it has destroyed their interference pattern. One may therefore attempt to make that measurement so that the electrons are affected only very, very slightly. Could one then have electrons that go through specific slits *and* also show the interference pattern?

In Section 1a we raised the question of how one can ever observe something without affecting it just a little bit. There, we consider fireflies; here, we deal with electrons. To locate one we must interact with it. In order to disturb the electron as little as possible we can try to use a very weak light source. The light reflected on the electron could be detected by very sensitive detectors and would then tell us through which slit the electron moves.

[handwritten left margin: cannot observe something w/o impacting it somewhat]

A low-intensity light source means, however, that very little energy is emitted per unit time. From the energy formula $E = nhf$ which simply says that there are n photons each with energy hf, we learn that a source can be made weak in two different ways: very few photons (n very small), and each photon with as low a frequency as possible (f very small). The lower the energy of each photon the less is the impact on the hit electron; its motion will then be less perturbed.

But neither of these reductions can be made with impunity. The reduction in the number of photons will eventually lead to a situation where there are not enough photons to observe (reflect on) every electron. Some electrons will come through the baffle without reflecting a photon and we won't know through which slit they passed. There must be enough photons around so that every electron will be 'seen' by a photon. We cannot reduce the number of photons *n* too much.

Similarly, we cannot reduce the frequency of the incident light *f* too much. As the frequency is reduced the wave length increases and the accuracy with which we can determine the location of the electron becomes worse: it is well known that every optical instrument has a resolution, i.e. a measure that tells how far apart two points must be so that they are actually seen as two points rather than as only one. And that resolution becomes worse as the wave length becomes larger. When one increases the wave length (decreases the frequency) two neighboring points appear more and more blurred until, when the wave length is about as large as the separation between the two points, they look like one single blur; they can no longer be resolved. In our case, the wave length must remain smaller than the distance between the two slits. Therefore, we cannot increase the wave length (decrease the frequency) of the light that we use to determine the position of the electron too much.

if decrease number of photons or frequency of wave too much, unable to observe electrons accurately

Our attempt at observing the electrons without disturbing them fails; a certain small amount of perturbation cannot be excluded. A careful study shows that, in fact, it can *never* be excluded. And the magnitudes of these disturbances are such that in our thought experiment we _either_ see which slits they go through _or_ we see the interference pattern. *— classical — quantum mechanics*

This alternative lies at the basis of quantum phenomena. It is an *intrinsic limitation*. It is indeed in the nature of things. And it translates in our case into: either we localize the electrons (determine which slit they go through) and find that they behave classically, or we don't and find they behave quantum mechanically. In the latter case they show interference [10.10].

In its mathematical formulation this intrinsic limitation is known as the *Heisenberg uncertainty relation*. One can state it as follows. Let us call *x* the distance along the baffle as measured from the bottom of the lower slit on up. The incident electron beam is so wide that the electrons can go through either slit. Any one electron will therefore have a probability distribution with a width Δx (for a bell-shaped curve this would be the width at half the maximum height). Δx is called the *uncertainty* of the probability distribution. In order to be sure that the electron passes through only one of the two slits, this Δx must be narrower than the distance *d* between them, $\Delta x < d$. On the other hand, the interference pattern is due to the electron's motion partly up or down after passing the slit, rather than going straight ahead. It depends on the component of its momentum p_x in the *x*-direction (or on its velocity in that direction since $p = mv$). The probability spread in momentum in that direction, Δp_x, must therefore be small enough not to wash out that pattern. Obeying the requirement for both, Δx and Δp_x, one finds that the product

$\Delta x \Delta p_x$ must be smaller than Planck's constant h.

The argument is as follows. The very small angle that the electron makes upward or downward is typically of the size λ/d. Therefore p_x is about equal to $\lambda p/d$. The probability spread Δp_x must be less than that, $\Delta p_x < \lambda p/d$. But $\lambda p = h$ according to the de Broglie relation. We must thus require $\Delta p_x < h/d$. If we take the product of the width Δx of the probability distribution in x ($\Delta x < d$) with the width in p_x that we need in order to have both localization and diffraction we find $\Delta x \Delta p_x < h$.

The Heisenberg uncertainty relation, one of the fundamental relations of quantum mechanics, however, tells us that this requirement can *never* be satisfied and that one always has

$$\Delta x \Delta p \geq h.$$

The product of the two uncertainties, one in the position and the other in the momentum, is always larger or, at best, about the same as Planck's constant (Fig. 10.6). Neither of these uncertainties can be completely absent. If one of them were zero (the position, say) the other one would have to be infinite (the momentum would be completely undetermined – an infinite uncertainty). No limits could be specified within which the value of the latter lies.

Applied to our thought experiment, the uncertainty relation tells us that the interference pattern goes hand in hand with a Δx large enough so that one cannot tell through which slit each electron goes. Any measurement of

Fig. 10.6. The uncertainty relation. For a given uncertainty in position, Δx, only those uncertainties in momentum, Δp, are allowed which yield points in the allowed region. At best, one can be *on* the curve. But even in that case, for small Δx one has to contend with a large Δp (point A), and for small Δp one has necessarily a large Δx (point B).

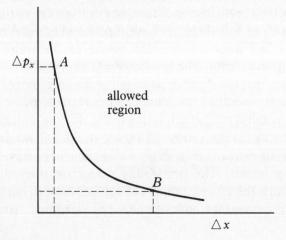

position, i.e. any attempt at localizing the electrons at one or the other slit will make Δx smaller, i.e. will narrow the probability distribution. But then Δp_x will *necessarily* become wider and that destroys the interference pattern. Thus, measurements can change the uncertainties but they cannot get around the limitations imposed on them by the uncertainty relation.

The uncertainty relation is entirely foreign to classical science. An object could have different properties at the same time, for instance a precise position *and* a precise momentum (or velocity). In the quantum world this is no longer possible as a matter of principle. There are also other pairs of properties for which such an uncertainty relation holds. But not all pairs of properties are restricted in this way. For example, the velocity components in two different arbitrary directions are not so restricted. This will be a subject of further study in Section 11a.

both position and velocity cannot be known (intrinsically)

Since h is a universal constant it establishes a *universal* lower limit on these uncertainties. And since this relation is characteristic of quantum mechanics (for classical mechanics the right side would be zero so that no restriction results) we can hold the constant h responsible for this quantum mechanical behavior. In a certain sense, and we shall discuss that in greater length in Section 10f, this quantum mechanical formula would become a formula valid in classical mechanics if h became effectively zero. Thus, h plays as fundamental a role in quantum mechanics as the speed of light c plays in relativity.

h

The mathematical formalism of quantum mechanics has the limitations imposed by the uncertainty principle as a *necessary* consequence. If anyone should find a way of getting around these limitations he would prove quantum mechanics to be violated. This has not happened in the more than 60 years since its inception and, judging from the tremendous success of that theory, it is very unlikely to happen.

The significance of the uncertainty relation lies in its ontic character. This means that it tells us something about the way in which quantum particles exist. They are 'spread out' in both their position and their momentum in accordance with that relation. But the uncertainty relation has nothing whatever to do with our ability to obtain a precise value in each single measurement. Only when the same measurement is repeated, will a different value be obtained, so that after a large number of measurements (of position, say) an overall uncertainty of Δx emerges. The sizes of Δx and of Δp_x depend on the state of the system and are always subject to the limitation of the uncertainty relation.

e. *Complementarity*

It happens quite often in the classical world of everyday life that we want to have as much information as possible about an object, and that it is impossible to gain all of it at the same time: obtaining some knowledge about it excludes the possibility of obtaining *simultaneously* certain other knowledge about it. An example will illustrate the point.

You want to purchase an expensive rug. You like the rug because of the artistic qualities you see, the design, the choice of colors, etc. This knowledge about the rug is acquired by looking at it from a certain distance so that the rug can be perceived as a whole. But there is other information about that rug which you must have in order to be able to decide whether it is worth the price. You also want to know how well the rug is made, how fine a weave it has, how it is knotted or woven, etc. For this purpose you must abandon your previous stance and inspect one small part of that rug very closely. It is impossible to make this inspection and at the same time obtain the overall view that permits you to judge its artistic appearance. Both types of information are necessary but they cannot be obtained simultaneously.

We call two kinds of information of this sort *complementary* because neither one by itself suffices to provide full information. They complement one another. Complete information may consist of a number of complementary pieces.

In the classical world of science the situation is quite similar. In studying how baseballs move one wants to know the ball's position as well as its velocity. In general, two different measurements are necessary for that. Both pieces of information about the ball can be obtained to as great an accuracy as one wishes. Both are needed for a complete knowledge of the state of the motion of the ball.

In the quantum world the situation is complicated by the uncertainty relation. The very nature of a quantum particle precludes its *simultaneous existence* in a precise position and with a precise momentum (or velocity [10.11]), and therefore it also precludes the corresponding simultaneous acquisition of knowledge. Nevertheless, both pieces of information are necessary for a complete understanding of that quantum particle; they complement one another.

To dramatize the situation, assume that a rug were a quantum object so that the accuracies of determining its design and its weave were related by the uncertainty relation. Inspecting its design puts the rug into a state of very uncertain weave, and inspecting its weave – into a state of very uncertain design. It's not that we just cannot *know* with greater certainty, but the object *exists* in a state of large uncertainty. Each experiment (inspection) puts the quantum object into a different state due to the interaction of the inspection apparatus with it. We shall return to that strange fact in Section 11b.

As we shall see in Section 11b the uncertainty relation is not restricted to position and momentum. Other pairs of properties may satisfy that relation. For each pair the two properties are complementary to one another. They are

both needed for a full description of the quantum system but they <u>cannot
have simultaneously sharp values</u> [10.11].

One can think of complementarity as an epistemic concept (related to our
knowledge of the object). But it has its origin in the ontic uncertainty
relation, intrinsic to the very being of the quantum objects.

f. *The essential link*

Enough has been said about quantum mechanics to leave no doubt that it
required a scientific revolution. Such diverse concepts as 'particle' and 'wave'
lost their classical meaning in the quantum world, and such old and cherished
classical concepts as the exact position and the exact velocity of an object
must be revised drastically. Classical determinacy is replaced by probabilistic
predictions, and measurements can no longer be made without affecting the
system to be measured. The uncertainty relation controls the situation.

SUMMARY of QM.

But we found in Chapter 8 that there cannot be a sharp border between an
old and a new theory. In ᵛlativity, for example, the old theory did not
become wrong and defunct. Furthermore, there are various necessary
conditions which such scientific revolutions must satisfy. The relations
between the superseding and the superseded theory must be spelled out; they
lead to the important notion of validity limits for that superseded theory. In
short, the link between the new level and the old level on which two scientific
theories operate is essential.

What is the relation between the quantum world and the classical world?
Surely, quantum mechanics cannot lay claim on the classical world: the
time-honored classical theories (Newtonian mechanics, classical electrodynamics, and all the rest) that were so successful for so large a variety of
physical phenomena cannot suddenly be dismissed as null and void. If not,
<u>where is the limit of the classical world</u> and where does the quantum world
begin? Where, indeed, are the validity limits of the classical world?

We owe the answer to these questions largely to Niels Bohr. He stated his
correspondence principle as early as 1918 and he added much to it later on.
There exists a certain correspondence between quantities in the quantum
theory and in the classical theory. As one goes from the microscopic to the
macroscopic domain, as the quantum nature of matter and radiation fades
into its large-scale classical picture, these quantities relate to one another. But
beyond that all those features of the classical theory which do not contradict
the quantum nature *are maintained* in quantum mechanics. For example, all
conservation laws continue to hold exactly (not just in a probabilistic sense!).

The quantum structure is of course characterized by Planck's famous
constant h, as we have seen repeatedly. But it would be incorrect to state the
transition from the quantum to the classical world as the limit in which h

becomes smaller and smaller and vanishes eventually. h is a quantity with dimensions (those of an angular momentum); that means it is not a number but it depends on the units we use (meter, second, kilogram). We encounter here exactly the same situation that we had in Chapter 8 in connection with the speed of light c. The validity limits of Newtonian mechanics had to be characterized by the *dimensionless* quantity v/c or, more precisely $(v/c)^2$. Similarly, the transition from the quantum to the classical world requires a dimensionless quantity, namely *the number of quanta that are involved*. If the energy of electromagnetic radiation is large so that it involves a large number of quanta of radiation (photons) then that radiation is well approximated by the classical theory. If the angular momentum L of an electron (see [9.5]) is very large because the quantum number n is very large, then L can be described in excellent approximation by the classical theory.

The transition from the quantum to the classical world can now be stated simply as *the limit in which the quantum numbers become very large*. Starting from the fundamental equation of quantum mechanics, the Schroedinger equation, one can in fact derive Newton's force law, $F = ma$, the fundamental equation of Newtonian mechanics in this limit. And there are many other equations of quantum mechanics that lead to corresponding ones of classical mechanics in the limit of large quantum numbers.

The validity limits of the classical theory are therefore characterized by the quantum numbers: when the quantum numbers become too small the quantum theory must take over. Otherwise, the errors made by the classical theory increase beyond tolerable limits. But let us now look at some specific examples.

Max Planck's black-body radiation law can be derived from quantum mechanical considerations. The curves (see Fig. 9.3) depend on the energy hf of photons of frequency f. If E is the total radiation energy of frequency f and if n is the number of photons of that frequency, then instead of hf one can write E/n. The limit to the classical radiation law can now be obtained when n is made very large. One obtains exactly the classical results of Rayleigh. Mathematically, there is no difference between making n very large (holding E fixed) or making E very small (holding n fixed); in both cases E/n approaches zero. Therefore, the classical result is obtained even when n is not large in the limit as E becomes very small. But E very small (n fixed) means very low frequency f. Therefore, the classical result can be obtained in the limit of low frequency (long wave length). And it was indeed for very long wave lengths that the classical result was found to agree with experiment (see Fig. 9.3).

Our second example resolves the apparent contradiction that exists concerning angular momentum. In classical mechanics it is a quantity that can vary continuously and smoothly; in quantum mechanics it is 'quantized': only changes of angular momentum in units of about the size of Planck's constant *h* are allowed. How can these be reconciled?

angular momentum not continuous [handwritten margin note]

The answer is quite simple. A jump from an angular momentum $nh/2\pi$ to $(n + 1)h/2\pi$ and from there to $(n + 2)h/2\pi$, etc., will not be seen as a discontinuous transition if *n* is very large. In that case the jumps are too small to be noticed macroscopically and the whole transition appears to be smooth.

The essential link between the quantum world and the classical world also requires an answer to a question that has surely occurred to the alert reader at an earlier time when we first distinguished between quantum particles and classical particles (Section 10b). Do all particles belong to one of these two classes, or is it possible that *under suitable conditions* a classical particle can act like a quantum particle and *vice versa*? More specifically, one can ask 'why don't tennis balls show interference on a suitable double slit baffle?'

The general answer to this type of question is that it all depends on the relative size of things. The de Broglie wave length of a typical tennis ball in motion is more than a trillion trillion times smaller than an atom. There are obviously no baffles with slits of comparable size, and even if there were, the ball is much, much too large to get through them. While this is an extreme example one can easily interpolate between *it* and the case of an electron. The electron's mass is so much smaller than that of a ball that even the larger speed (a million times faster than the ball's, say) would not prevent its momentum (which is mass times speed) from being a trillion trillion times smaller. And that entails a de Broglie wave length that is that much larger, i.e. of approximately atomic size.

If we start with an electron with its interference pattern from a double slit (Fig. 10.3(*a*)) and gradually increase its mass (a thought experiment) keeping the speed and the baffle unchanged, we will find a smaller and smaller wave length. Correspondingly, the interference pattern would still cover about the same area of the screen but would have its maxima squeezed together more and more. Soon these maxima can no longer be resolved and they appear as a uniform average value. After enough particles have come through the slits so that high enough statistics have been accumulated, the pattern will appear as a smooth distribution. The effect is still there but it is completely hidden and our perception of what there is has completely changed.

As an example of a change in perception consider the reproduction of a photograph in a newspaper. It is printed as a very large number of tiny dots (a dot matrix) which gives the impression of a uniform distribution that changes gradually to show the features of the picture. In poor-quality printing this dot matrix is easily visible but the eye ignores it when one tries to perceive the picture. In high-quality printing the eye is fooled; the dot matrix is no longer visible. Only the picture is perceived.

The conclusion is that the distinction between quantum particles and classical particles is one of circumstances. There are objects in nature which we call particles (including tennis balls and electrons). Whether they behave classically or quantum mechanically depends on their properties (mass, size, speed, etc.), the circumstances under which they are observed, and the particular kind of observation that is being carried out on them. Under suitable conditions they may behave in either way. But in very many cases one or the other set of conditions cannot be realized for a given particle, so that some particles can never behave quantum mechanically and others only seldom classically. Therefore, we might call them 'classical particles' or 'quantum particles' as the case may be.

A different example of the importance of relative size is provided by the contrast between Compton scattering of X-rays by electrons (a quantum mechanical phenomenon – see Section 9b) and the scattering of an electromagnetic wave of low frequency on electrons (a classical phenomenon). In both cases one deals with radiation but its description is entirely different. The crucial relative size here is the ratio of the energy of the photon to the rest energy of the electron, hf/mc^2. When that ratio is about 1 or larger than 1 (as is the case for energetic X-rays) the electron receives energy and momentum to recoil. One has the Compton situation of the collision between two particles, the photon and the electron. The scattered photon then has necessarily less energy than the incident one and therefore has lower frequency.

But when the ratio is much less than 1 (sufficiently low f) the photon is too weak to budge the heavier electron. There will be no recoil. The photon will then not lose energy but it will be deviated in its path. If only very few photons at a time hit the electron one can detect them scattered in all directions. With a large enough number of photons in the radiation one finds the *classically* predicted distribution of radiation scattered by electrons: since there are very many photons they will act effectively as an electromagnetic wave precisely in the way classical electrodynamics describes it. Individual photons will no longer be detectable. The distribution in direction of the scattered radiation will be exactly the classically predicted distribution. The 'graininess' of the electromagnetic wave is no longer perceived.

This answers a question that was raised in Section 9b: radiation can be an electromagnetic wave (as classical electrodynamics will have it) or a stream of photons (as quantum mechanics sees it); it all depends on the circumstances.

The phrase 'under suitable condition' thus controls whether 'particles behave like waves' or whether 'waves behave like particles', using the naive terminology. It refers to the relative size of the quantities that determine the characteristics of the phenomena (distances, energies, speeds, etc.). That size decides whether a physical system is correctly described within the quantum world or within the classical world.

Annotated reading list for chapter 10

Bohr, N. 1934. *Atomic Theory and the Description of Nature.* Cambridge: University Press. Also 1963. *Atomic Physics and Human Knowledge.* New York: Wiley. These are the best sources for the interpretation of quantum mechanics as seen by the senior founder of that theory.

Feynman, R. P., Leighton, R. B., and Sands, M. 1965. *The Feynman Lectures on Physics.* Reading: Addison-Wesley. This is one of the best and most sophisticated introductions into physics on the college level. The first few chapters of the third volume are not very mathematical and give an excellent discussion of the double-slit experiment.

Jammer, M. 1966. See the annotated reading list for Chapter 9.

Pagels, H. R. 1982. *The Cosmic Code.* Toronto: Bantam Books. A popular and lively introduction into quantum mechanics and particle physics by a competent theoretical physicist.

Pais, A. 1986. See the annotated reading list for Chapter 9.

11

From apparent paradox to a new reality

This chapter deals with the heart of the conceptual revolution brought about by quantum mechanics. The situation that led to this revolution can be stated most simply in terms of two facts: (1) After some 60 years of existence quantum mechanics can be judged to have been extremely successful and to have become an established theory (in the sense of Chapter 8) with well-known validity limits and no negative empirical evidence within those limits. (2) Quantum mechanics has a well defined and sophisticated mathematical structure on which all its successful computations are based.

Given these two facts one now asks what this theory tells us about the nature of the quantum world. What model of that world is described by those mathematical symbols that work so well? What mechanisms are at work on the atomic and subatomic scale? Where does the mathematics of quantum mechanics lead us? In short, we are seeking an *interpretation* of that abstract theory. Since it has been so successful both in accounting for known phenomena and in predicting new ones (both qualitatively and quantitatively), great importance is attached to such an interpretation.

Without an interpretation quantum mechanics would be an empty computational scheme devoid of meaning. In a strict sense, it could not even lay claim to providing explanations.

But exactly at the point when one attempts to provide for an interpretation, difficulties arise. Naive attempts of this sort lead immediately to paradoxes. These paradoxes tell us that the theory 'makes no sense' and offends our sensibilities. The early problem of the dualism of wave and particle is just one example. One is forced to regard the quantum world that emerges from quantum mechanics as quite different from the world we are used to (the classical world) and in which our common sense has been trained. We must give up our classical views and be guided by what quantum mechanics tries to tell us about nature.

The results of such efforts have been most rewarding. There emerges a new view of reality which holds in the quantum world, and which is indeed quite different from that of the classical world we know so well. It is this exciting adventure, this discovery of a new view of reality, that rewards the persistent seeker of truth. That is the story that will be told in the present chapter.

a. *Quantum systems*

We shall use the term 'quantum system' for the objects of atomic and subatomic dimension which are described by quantum mechanics. Later on we shall see that even macroscopic systems can be quantum systems because their behavior is controlled by quantum mechanical effects (Section 11g). Examples of quantum systems include: fundamental particles such as electrons and protons, atoms, molecules, pairs of interacting particles (e.g. an electron colliding with an atom), and unstable particles that decay into other particles.

Since we already know that a probabilistic description plays an important role for quantum systems it will be helpful to consider first classical macroscopic systems that we know well and that are also of a probabilistic nature. A pair of dice is a good example.

A die is an exact cube and is of uniform density when true, that is when it is not 'weighted'. When thrown, the probability of showing any one side after coming to rest on a table is exactly one in six. All six sides of a true die must have equal probabilities for the outcome of a throw. In order to test that the die is true we can make a very large number of throws and record the outcomes. As the number of throws increases the relative frequency of showing any one side must approach the ratio $\frac{1}{6}$. Of course, it would require an infinite number of throws to reach this limit exactly. But a fairly large number of throws will give a sufficiently good test for the trueness of the die. Unfortunately, such *a relative frequency test* is often possible only as a thought experiment. In many probabilistic situations one cannot repeat the same situation, or one can at best make only a very limited number of repetitions. Just think of the prediction: there is a 30 per cent chance of rain tomorrow. Therefore it is often necessary to accept such initial (or *a priori*) probabilities which cannot be tested by a relative frequency test. This is also the case in quantum mechanics.

If we throw a single die and bet on the side with the five dots to be the outcome we have a chance of one in six to win. Let us now first concentrate on the time interval after the die has been tossed but before the die has come to rest on the table. The die is tumbling through the air in an unpredictable fashion. Of course, we are told that physicists know how to compute the motion of such a body as it moves through a gas (air) provided they know the initial position and velocity of the die, its initial spin, the density and temperature of the air, etc. We all believe that *determinism* also applies to this particular system but we also know that the necessary data are in practice not available. This is the reason why playing dice is a game of chance. While the system is intrinsically deterministic, our *knowledge of it is only probabilistic. We have ontic determinism but epistemic indeterminism.*

Let us make a closer inspection of this epistemic indeterminism. While the die is still tumbling through the air, we can ascribe to it a 'potentiality' or 'propensity' to come up with a 5 eventually. This potentiality is measured by the probability of $\frac{1}{6}$. It is present just because the die is in that indeterminate state. After the die comes to rest that potentiality turns into an 'actuality', the probability into a certainty.

Our bet on the outcome 5 with its chance of $\frac{1}{6}$ suffers a drastic change when the die comes to rest. We either win or lose depending on whether it shows 5 or does not show 5. As it comes to rest a sudden change takes place in the probability of showing 5: it changes from $\frac{1}{6}$ to 1, or from $\frac{1}{6}$ to 0. For later use in quantum mechanics we shall call this sudden change of probability a 'collapse' of probability to certainty. It is the collapse of a potentiality to an actuality.

There is nothing mysterious about this collapse. It is characteristic of any probabilistic situation when the outcome becomes known.

Let us now turn to quantum systems. One quantum system we know well is an electron going through a two-slit baffle and then hitting a screen (Section 10a). Quantum mechanics describes that electron by the probability amplitude ψ. This tells us the potentialities of that electron: where on the screen it will more likely hit and where less likely. As the electron hits the screen those potentialities turn into an actuality: one of the potentialities is being realized (actualized). This realization is ascertained by the observation of a blip at that spot on the screen.

Consider now another quantum system, in fact a much simpler one than the double-slit arrangement: a sample of radioactive radium that emits an alpha particle spontaneously and thereby changes into the element radon. The ψ which describes that decay is a spherically symmetric probability wave. It moves outward from the nucleus in all directions at the same time like an expanding balloon, and it tells us that the chance for that alpha particle to be emitted in any one direction is exactly as large as in any other direction. When it eventually *is* emitted it goes in one particular direction, the ψ collapses, and the potentiality becomes an actuality. One can surround the radium sample by a sphere of very many detectors. Any one of these may respond and record an alpha particle; but only one of them does, indicating the arrival of an alpha particle from the radium sample in the center. The ψ gave no such account. It only gave the potentialities and not the actuality.

This indeterminism is intrinsic to the quantum system, it is an *ontic indeterminism* in contradistinction to the classical case.

Not everyone accepted this interpretation of quantum mechanics. Albert Einstein expressed his opposition most eloquently by his famous remark 'God does not play dice'. He remained faithful to classical determinism. We shall consider his objections to quantum mechanics in Section 11d.

Another interesting probabilistic property of the decay of a radioactive substance is the following: the time at which it emits a particle cannot be predicted. That time is governed only by probabilistic laws. What we do know is that for each such substance there is a certain characteristic time period T called the 'halflife'. It is the time it takes for exactly half of the substance to decay. For example, for naturally occurring radium $T = 1620$ years, while for radioactive cobalt (used in medical therapy) $T = 5.24$ years. After time T half of the atoms in the sample will have decayed; after another period T half of the remaining atoms will be gone so that there are only one quarter of the original atoms left, etc.

The halflife is an average which can be computed from ψ. But there is no way of telling exactly when any one particular nucleus would decay. A similar situation exists when insurance companies have mortality tables that tell the probability for a person to live another 20 years. They cannot tell the life expectancy of any one individual; they can only give him probabilities. Here, too, we have only probabilities. The actual future is not predicted.

Thus, in summary, the probability amplitude ψ of a quantum system provides a 'list of potentialities' for the quantum system. But contrary to classical systems, there is no underlying determinism.

b. *Observables and measurements*

Physical properties include such things as size, speed, weight, electric charge, energy, frequency, etc. All these quantities have a certain magnitude which can be measured. The measurement yields a number that tells us how large a particular physical property is in terms of the units we have chosen. Thus, when measuring the length of a rod we use a meter stick (the chosen unit) and find how many times longer (or shorter) the rod is.

This 'classical' notion of a measurement takes for granted what are actually two assumptions. The first is that the physical properties to be measured already *have a definite value before the measurement* is made. The second assumption is that *the measurement does not affect the object* and does not change the value of the property to be measured. It gives a faithful rendition of what there is before the measurement and leaves it unchanged. To be sure, there are exceptions to it. A case where the observation influences the observed system has already been considered in Section 1a in the context of a non-quantum mechanical situation: seeing a firefly (Section 1a). But in the large majority of the usual measurements in the classical world it is justly assumed that no such influence occurs.

In the quantum world neither of the above assumptions is valid. How they are violated is the topic of the present section.

In the language of quantum mechanics physical properties are called *observables* and one often speaks of 'the observer' in discussing meaurements. This is a very unfortunate use of words. It has historical reasons which we do not need to pursue here. But these words tend to imply that the observer as a person is somehow involved in quantum mechanical measurements. This is absolutely not the case, certainly not more so than in classical measurements: in both cases the observer is of course responsible for building the measuring apparatus. But in every other sense quantum mechanical measurements are just as objective as classical measurements. They are what scientific measurements should be.

The simplest way to eliminate the observer as a person is to insist that all measurements must leave a permanent record. Examples would be a record on a photographic film or on a tape recorder; or it could simply be an automatic recording of the pointer position of a measuring instrument. In this way the measurement does not depend on the observer's presence at the time of the measurement. It can be read off by anyone at any time.

What does happen in the quantum world and does not usually in the classical world is this: *the measuring apparatus in general changes the observed object (the 'system') during the measurement*. The second assumption above (that the measurement does not affect the measured object) is therefore false although there are exceptional cases. In the following we shall have a lot more to say about this *system–apparatus interaction*.

The first assumption is also false: many observables in the quantum world do not have a definite value before the measurement is made. The situation is somewhat like that of a die while it is still tumbling through the air: it shows no definite number. What does exist is the *potentiality* (see the preceding section) to take on one of several different values just like the tumbling die does. The latter can take on six different values (in the sense that six different surfaces may show when it comes to rest). Quantum mechanical properties (observables) can take on a finite number of values or an infinite number of them. The property is in a sense distributed over all these possible values, it is *blurred*. In the classical world, by contrast, properties have a definite value; they are *sharp*.

A good example of a blurred property is the position of an electron inside an atom. The electron is a point particle [11.1] so that, classically, one would expect it to be at a specific point in the atom at a given instant. The electron's position is expected to be sharp. This is not the case. The electron is distributed all over; it is spread out all around the atomic nucleus like a cloud; it is blurred. It has a probability distribution which tells us where it is more likely to be found and where less likely. This probability distribution is not 'just a matter of our ignorance'. It is intrinsic to the state of the electron inside an atom. That's the way the electron actually *is* when bound to the nucleus. The blurring is an ontic feature.

In the classical world all observables are *compatible*: they can all be sharp at the same time. This means that a body can have *simultaneously* an exact position, an exact velocity, an exact energy, etc. Correspondingly, measurements can be carried out that permit the simultaneous determination of all these observables to arbitrary precision restricted only by our ingenuity in devising clever measuring techniques.

In the quantum world this is not the case. Given any one observable, O, all the others fall into two classes, those *compatible* with O and those *incompatible* with it. There is a simple and precise mathematical test in quantum mechanics that permits one to decide whether two observables are compatible with one another or not. No ambiguity whatever exists about that. Any observables compatible with O can be sharp at the same time as O. But none of the incompatible ones can.

We have already encountered a pair of incompatible observables: the position and the momentum of a quantum particle. The uncertainties Δx and Δp_x (Section 10d) are a measure of their blurring. And the uncertainty relation tells us that they cannot both be sharp at the same time.

When a measurement of an observable O is made, the first question is whether the quantum system (object) before the measurement is in a state in which O is sharp. If this is the case, the measurement will reproduce the value that O had before the measurement. But if O was not sharp before the measurement, the result will be one of the possible values that O can take on. One of its potentialities will be actualized. The blurred observable will be made sharp due to the interaction with the apparatus. This is in essence what the apparatus does to the system during a measurement.

The measuring process in the quantum world is here seen to be much more complicated than the measuring process in the classical world. In the latter the measuring instrument, the apparatus, is idealized in that it is assumed not to affect the system to be measured. Classically, this is indeed an excellent approximation in most instances. But quantum mechanics is much more realistic in that respect. It treats the apparatus like any other physical system in that it takes full account of the interaction that always takes place between it and the system to be measured. In quantum mechanics the apparatus *can* affect the system, and it always does so when an observable is measured that is not sharp.

According to quantum mechanics the physical state in which a particular quantum system exists is completely specified by all the compatible observables, and only by these. The incompatible ones are unimportant for knowing all there is to know about that state. The number of compatible observables necessary to know the *state* of a quantum system completely is called a *complete set of compatible observables*.

This notion of completeness leaves out a lot of observables that would be compatible on the classical level. The quantum system is therefore called 'complete' with many fewer data than a classical system would require [11.2].

The claim to completeness seems to be in tension with the principle of complementarity (Section 10e). The use of only compatible observables and the need for the complementary (incompatible) observables to complete the description of the system seem to produce an apparent paradox. The resolution of that paradox is not difficult. It is due to confusion about the complete knowledge of the state of a system and of the system itself. For example, a system with sharp energy (such as an atom with lowest possible energy) can be in many different *states* depending on which other observables are sharp together with the energy [11.3]. All these states complement one another; they are all needed to know the *system* completely.

The matter can be understood by a simple analogy to the macroscopic world: one can view a piece of sculpture (the analogue to our system) in good light and very clearly from one particular angle. (That view is the analogue of the state of our system.) But the same clear view from a different angle will complement our knowledge of that sculpture. In fact, one must view it from several different angles to obtain a full visual appreciation. (We ignore here non-visual appreciation by touching and feeling.) The different views complement one another.

In any case, we can see now that there is no conflict between completeness and complementarity. Complementary variables such as position x and momentum p cannot be sharp simultaneously. Correspondingly, two states of the same system which differ in that x is sharp in one and p is sharp in the other cannot exist at the same time. But the information each of them contains complements the other.

Let us now consider an example of a quantum system in a given state and the measurement of an observable incompatible with the observables characterizing that state. One interesting property of electrons is their *spin*: electrons spin (like a top), and the amount and sense of spin rotation is given by an observable called *spin angular momentum*. According to quantum mechanics, all electrons have at all times the same amount of spin angular momentum. It is $\frac{1}{2}\sqrt{3}$ angular momentum units (an angular momentum unit is Planck's constant h divided by 2π). But if one chooses any line through the electron and measures the component of spin along that line (its projection onto that line) one finds a value of $\frac{1}{2}$ angular momentum units. Since that component can point in the positive or the negative direction along that line, one finds either $+\frac{1}{2}$ or $-\frac{1}{2}$. This is the complete set of values the spin can have along a given line.

If the electron before the measurement is in a state with the observable 'spin along the x-direction' (call it s_x) of value $+\frac{1}{2}$ then the spin component s_y along some other direction (call it the y-direction) is blurred because s_y and s_x are incompatible observables. This incompatibility is a mathematical consequence of the theory.

From the probability amplitude ψ one can, however, compute the probability P_+ that a measurement of s_y will yield the value $+\frac{1}{2}$. Similarly, the probability for finding $-\frac{1}{2}$, P_-, can be found. Since one of these two values for s_y *must* be the result, these two probabilities add up to 1 (which indicates certainty), $P_+ + P_- = 1$. The actual values of P_+ and P_- depend on the angle between the two directions, the x-direction and the y-direction. If that angle is a right-angle then the two probabilities are equal, $P_+ = P_- = \frac{1}{2}$.

One can express this situation also by saying that the original ψ which describes a state with the sharp value $+\frac{1}{2}$ for the observable s_x also tells us the potentialities available for the outcome of the measurement of the incompatible observable s_y. But it does not tell us which of these potentialities will actualize. Quantum mechanics cannot predict the result of the measurement of an incompatible observable. For such a prediction the theory is intrinsically probabilistic. This means that the theory *does* predict precisely the probabilities P_+ and P_- for getting one or the other result. One can test that by means of the relative frequency test. One simply repeats the same experiment over and over again; each time one must start with an electron that has $s_x = +\frac{1}{2}$, and each time one must measure s_y. After a very large number of repetitions the fraction of times the result $+\frac{1}{2}$ is found should be P_+, and the fraction of times the result $-\frac{1}{2}$ is found should be P_-.

What happens therefore when one makes a measurement of an observable that is incompatible with the ones that are sharp in the system to be measured? The system is changed from its original state to a different state. The new state is characterized by a different complete set of compatible observables which are all simultaneously sharp. That set now includes the observable that has been measured. In the above example the original state has sharp observables which include s_x while the new state has sharp observables which include s_y. All observables characterizing the new state must of course be compatible with s_y as well as with one another.

This change of state of the system can be thought of in terms of the analogy which we used earlier. Measuring two different spin components of the electron is like looking at a sculpture from two different sides. Just as the sculpture looks different from two different sides, the electron is found to be in different states when measurements are made for its spin along different directions. The difference between a quantum system (the electron) and a classical system (the sculpture) is this: the two states of the electron are incompatible (corresponding to the incompatible observables s_x and s_y); the system is therefore either in one state or the other but it cannot be in both at the same time. The two views of the sculpture, however, are compatible; the sculpture does not change when we change our angle of view.

c. *Schroedinger's cat*

One of the most outstanding physicists who objected to the interpretation of quantum mechanics developed by Niels Bohr, Heisenberg, and others, and which we have been following so far, was Erwin Schroedinger. His criticism was especially remarkable because he had been one of the founders of the quantum mechanical formalism. The fundamental equation of quantum mechanics which describes the development in time of the probability amplitude ψ carries his name.

In 1935, in the same year in which Einstein (in collaboration with others) also wrote a critical paper on the subject (see the following section), and after an interesting correspondence with him, Schroedinger published his own criticism of quantum mechanics in a very long paper. One of the many points he made was his claim that the interpretation of ψ as a probability amplitude does not make sense when applied to macroscopic objects and must therefore be discarded. His argument consisted of the following thought experiment (Fig. 11.1).

A completely closed chamber contains a live cat and a small amount of radioactive material. When a particle is emitted by one of the disintegrating atoms of that radioactive material, it triggers a detector that in turn releases a hammer which smashes a small flask containing a poisonous gas. This contraption is shielded from interference by the cat who dies almost instantly from that gas. It is a diabolic device.

Now the radioactive source behaves according to the laws of quantum mechanics: the time when a particle will be emitted cannot be predicted; one knows only the probability for this to happen. Let us assume that probability to be $\frac{1}{2}$ per hour which means that there is a fifty-fifty chance for a particle to be emitted during the next hour.

If we can consider the whole chamber, cat and contraption together, as 'the system' then, argues Schroedinger, it has a quantum mechanical probability amplitude ψ. And that probability amplitude would after one hour tell us that the cat is half-alive and half-dead. It is the sum of the ψ for the cat being alive and the ψ for the cat being dead. It states therefore that the cat is half-alive and half-dead (!). Only after an observer opens the chamber can it be ascertained whether the disintegration of one of the atoms has actually taken place by observing whether the cat is actually alive or dead.

What Schroedinger attempted to do in this thought experiment was to transfer the potentialities of a quantum mechanical system, a radioactive atom, to a macroscopic system, a cat. The atom has two potentialities, to disintegrate or not to disintegrate. This results in a ψ for the cat with the two potentialities of live and dead. And that, argued Schroedinger, does not make sense at all because a cat is always either alive or dead. Therefore, the interpretation given to ψ cannot be maintained.

Expressed in other words, a microscopic indeterminacy (when a particle would be emitted) has been turned into a macroscopic one. A macroscopic object, a cat, is claimed to be in an indeterminate state of two mutually exclusive possibilities, being alive or being dead. Schroedinger thus seems to have succeeded in showing the inadequacy of the theory. Or did he?

Fig. 11.1. Schroedinger's cat.

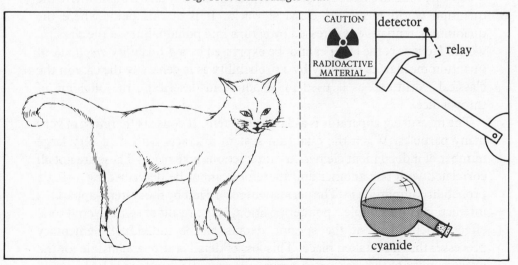

Schroedinger treated the chamber, cat and device, as one single system which is in a pure quantum-mechanical state. He assumed that the probability amplitudes for being alive and being dead can be added. But this is not possible as we shall now explain.

In his argument Schroedinger made use of a fundamental principle of quantum mechanics, the *superposition principle*. It says that the ψ for a blurred observable in a pure quantum state is the sum of ψs one for each of the potentialities of having a definite value [11.4]. Therefore,

$$\psi_{CAT} = \psi_{ALIVE} + \psi_{DEAD}.$$

But this principle is not applicable for a macroscopic object like a cat for the following reason:

In a sense, the cat acts as a measuring device which tells us whether or not the atom has decayed. That is of course as it should be: the apparatus, the cat, is a macroscopic object that provides a permanent record of the measurement. But there are other things taking place between the disintegrating atom and the dying cat. There is a chain of events between the two: the emitted particle triggers the particle detector, the particle detector trips the hammer, the hammer breaks the flask, the poisonous gas escapes, the cat breathes the gas. All these events are macroscopic classical events and can be correctly described classically except the first one. The important location in this chain of events where the quantum-mechanical phenomenon is 'amplified' into a classical phenomenon and where the classical description can take over is therefore the particle detector. That is the point where the first *permanent macroscopic record* of the event is made.

In every measurement of a quantum system there is such a point where the quantum world meets the classical world. It is at this point where the quantum potentialities expressed by ψ turn into potentialities of the classical world. However, the latter cannot be expressed by a probability *amplitude* of quantum mechanics but only by a probability as is generally the case in the classical world and as is used for example in describing the outcome of thrown dice.

The measuring apparatus is a classical system. It consists therefore of very many particles. When the quantum system interacts with it, a very large number of independent elementary interactions take place. These are not all correlated with one another and therefore cannot be described by a single probability amplitude ψ. The permanent record left by the system–apparatus interaction (for example a pointer position) is the result of some overall total effect independent of the specific details of the individual elementary processes that have taken place. This loss of detail is also responsible for the irreversibility of the measuring process; it cannot be undone [11.5]. In any case, there is a very clear demarcation in the long chain from the radioactive decay to the dead cat which specifies where the transition from a quantum system to the classical apparatus takes place.

Thus, the quantum mechanical potentialities of the radioactive atom (to disintegrate or not to disintegrate) have been turned into an actuality in the *detector*. What happens in the chain of events that follows the detector is in the domain of classical physics. The use of ψ for the description of the cat is therefore not justified. Schroedinger's case against quantum mechanics fails.

d. *Einstein's reality*

There are two simple laws of probability. Suppose we know that the event A will occur with probability P_A. Event A may be 'throwing a die with the outcome 5'; in this case P_A is $\frac{1}{6}$. Suppose we know that event B will occur with probability P_B. We assume that the two events A and B are independent of one another and that they are mutually exclusive: if one occurs the other one cannot occur. Then the first law is the following:

(1) The probability for either A or B to occur is $P_A + P_B$.

This is called the addition law of probabilities. The probability for the die to have the outcome 5 is $\frac{1}{6}$; the probability for the outcome 4 is also $\frac{1}{6}$. The probability that the outcome will be either 5 or 6 is $\frac{1}{6} + \frac{1}{6} = \frac{1}{3}$.

If the two events A and B *can* occur simultaneously (i.e. when they are not mutually exclusive but still independent) the second law states:

(2) The probability for both events to occur together is $P_A \times P_B$.

This is called the multiplication law of probabilities. The probability for the red die to yield 5 and in the same throw for the green die to yield 4 is $\frac{1}{6} \times \frac{1}{6} = \frac{1}{36}$.

If the probabilities P_A and P_B are not independent of one another then the multiplication law does not hold. In that case one speaks of a (classical) *correlation* between the events. If P_A is the probability that a person picked at random is over 70 years old, and if P_B is the probability that a person picked at random is sick, then the probability of picking a person who is both over 70 and sick is not $P_A \times P_B$ because there is a correlation between these events: older people are more likely to be sick than younger ones.

In quantum mechanics these laws hold also, but with the important change that the probability *amplitudes* (rather than the probabilities) are added and multiplied, respectively. The following example will demonstrate the matter:

Suppose two electrons A and B are produced together and ejected simultaneously from some source; and suppose also that they start out with their speeds along the same line but in opposite directions. Let us further assume that the whole two-electron system was produced without spin (see Section 11b). Since the spins of the two electrons along any line therefore add to zero and since spin angular momentum is conserved (just like energy), the two electrons will move away from that source (and from one another) in such a way that their spins will continue to be aligned in opposite directions.

There are two ways in which this can happen: either particle A has spin $+\frac{1}{2}$ and particle B has spin $-\frac{1}{2}$, or B has $+\frac{1}{2}$ and A has $-\frac{1}{2}$. If this were a classical system, the multiplication law of probability (item (2) above) would have to be used. But since this is a quantum system one must use the multiplication law for probability *amplitudes*. Therefore, in the first case the probability amplitude ψ_{AB} for the two-electron system is symbolically

$$\psi_{AB} = \psi_A(\tfrac{1}{2}) \times \psi_B(-\tfrac{1}{2}),$$

and in the second case it is

$$\psi_{BA} = \psi_B(\tfrac{1}{2}) \times \psi_A(-\tfrac{1}{2}).$$

There is no problem with either of these cases: when the probability is computed from the probability amplitude in each of these two cases (by computing the square as explained in Note 10.6) it yields just the product of the two probabilities exactly as the above multiplication law requires in the classical world;

$$P_{AB} = P_A \times P_B, \qquad P_{BA} = P_B \times P_A.$$

In each case the probability amplitude provides a complete description (in the quantum mechanical sense) of two independent (non-interacting) particles. Each particle is in fact associated with a complete set of compatible observables.

But now we realize that we do not have just the first case, nor do we have just the second case. Either the first or the second case can occur so that we should use the addition law of probabilities (item (1) above). But again, quantum mechanics requires us to add the probability *amplitudes* rather than the probabilities, resulting in the amplitude for the whole two-electron system,

$$\psi_{(\text{whole})} = \psi_{AB} + \psi_{BA}.$$

The addition of two ψs in this fashion to obtain a new ψ is just the *superposition principle* that Schroedinger applied incorrectly to a classical system (a cat – see the preceding section).

The probability amplitude for the whole two-electron system, $\psi_{(\text{whole})}$, is now no longer the product of a ψ_A for electron A times a ψ_B for electron B. It involves a sum of two (probability) amplitudes and thus can give rise to interference as in the double-slit experiment. It means that the two parts of the $\psi_{(\text{whole})}$ somehow 'know of one another'. This appears experimentally as a *correlation* which does not exist in the classical case and which is different from a classical correlation. It is called a *quantum correlation* and we must study its implications.

Because of this correlation, the $\psi_{\text{(whole)}}$ no longer describes each of the two electrons *completely*. We cannot tell from $\psi_{\text{(whole)}}$ whether electron A has spin $+\frac{1}{2}$ or $-\frac{1}{2}$, and similarly for electron B. $\psi_{\text{(whole)}}$ is no longer associated with a *complete* compatible set of observables either for electron A or for electron B. It *is*, however, associated with a complete compatible set of observables for the two electrons together, i.e. for the whole system. In the next section we shall raise the question whether this is an epistemic or an ontic phenomenon, i.e. whether this refers only to our knowledge about the system or whether the system really exists in this state.

But in either case, we have here a surprising result: the complete quantum description of the whole implies an only incomplete quantum description of its parts. And one can also show the converse: the complete description of the parts (the two electrons) does not provide a complete description of the whole.

The loss of some of the complete description of the parts (because they belong to a larger whole) was called *entanglement* by Erwin Schroedinger. It is the same as a *quantum correlation*. The two electrons which emerge from the atom are entangled (or quantum correlated) despite the fact that they do not interact with one another. Neither Einstein nor Schroedinger were ever willing to accept this notion.

But here, then, is a new concept: a system of two free particles, particles that are not exerting any forces whatever on one another, can come in two kinds in the quantum world: independent particles (non-entangled) and entangled ones. The latter kind does not exist in the classical world.

We were thus led into a view of the quantum world which is a kind of holism. This emergence of a holistic view is not just philosophical talk: one can make experiments to check on it and we shall discuss one of these in the following section, 11e. But before one would make such experiments one would construct thought experiments for it. That brings us to the master of thought experiments, Albert Einstein.

Einstein, who had contributed so much to the development of quantum mechanics starting with the light quantum as a concept to be taken seriously, stood firmly on the grounds of *classical realism*. A real object has real parts, and real things must also have complete descriptions (sharp observables) when they are only parts of a bigger system. Quantum mechanics, he felt, gives an *incomplete description* of nature; and so he devised various thought experiments to prove his point. The most famous of these was developed in collaboration with Boris Podolsky and Nathan Rosen, two other members of the Institute for Advanced Study in Princeton, Einstein's permanent location after leaving Nazi Germany. It is known as the Einstein–Podolsky–Rosen experiment or simply as the *EPR thought experiment*.

The original EPR thought experiment would be very difficult to turn into an actual experiment. Therefore a modification was suggested by David Bohm that *can* be carried out, and that has in fact been carried out in different laboratories (see the following section). We shall concentrate on that version; but first let us discuss it as a thought experiment.

Consider our two electrons A and B emerging simultaneously from some source. Suppose they are traveling in opposite directions. Let A be the one that travels to the right and B the one that travels to the left (Fig. 11.2). Let us also have detectors located to the right and to the left. These detectors are able to measure the component of spin of each electron along *any chosen direction* perpendicular to the motion of the electrons. They can tell whether the spin is up $(+\frac{1}{2})$ or down $(-\frac{1}{2})$ along any such line (call it the x-direction).

Thus, we are given the state of the whole two-electron system $\psi_{(\text{whole})}$ and we measure the spin of electron A in the x-direction. Suppose we find it to be up $(+\frac{1}{2})$. Then we can deduce from $\psi_{(\text{whole})}$ that we are dealing with ψ_{AB} rather than ψ_{BA}, and that therefore the spin of electron B along the x-direction must be down $(-\frac{1}{2})$. Similarly, if we find the A spin to be down, we infer that the B spin must be up (we have the state ψ_{BA}).

There is the possibility, as was indeed suspected at one time, that the successive production of these electron pairs occurs actually as mixture: once in the state ψ_{AB} and another time in the state ψ_{BA}, alternating randomly between these two possibilities. This is *not* the meaning of $\psi_{(\text{whole})}$, and an experimental check has verified it; they are not produced as such a mixture.

It follows as the first important observation according to quantum mechanics that a change occurs of the spin of each electron from a *blurred*

Fig. 11.2. The Einstein–Podolsky–Rosen thought experiment as modified by Bohm.

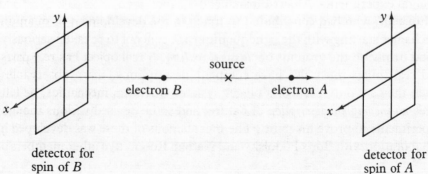

observable before the measurement (from the state $\psi_{(\text{whole})}$ in which each spin has an equal potentiality to be up or down) to a *sharp* observable afterwards. This change is brought about by the interaction of the two-electron system as a whole with the apparatus. It is an example showing how in quantum mechanics the apparatus can influence the object to be measured (Section 11b).

This change from the entangled to the disentangled state is called the *collapse of* $\psi_{(\text{whole})}$ to ψ_{AB} (or ψ_{BA}). It is the actualization of a potentiality.

We have encountered this notion before (Section 11a). In the present case, $\psi_{(\text{whole})}$ has the two potentialities described by ψ_{AB} and by ψ_{BA}. When the measurement of A yields the result that its spin is $+\frac{1}{2}$ then $\psi_{(\text{whole})}$ collapses to ψ_{AB}; and when it yields $-\frac{1}{2}$ it collapses to ψ_{BA}. In either case, *the entanglement is broken* and the resulting new system satisfies the product law, $P_{AB} = P_A \times P_B$, or $P_{BA} = P_B \times P_A$, as in the classical world. The measurement has successfully destroyed the entanglement (technically: the quantum correlation) during the time that the system and the apparatus interacted. After that time, the measurement is completed, the entanglement is broken, and the two particles are in a new state. For this new state one has a complete description of each particle, i.e. a complete set of sharp compatible observables for each.

The second important observation according to quantum mechanics is that we can obtain a prediction about the sharp value of the spin of B – which is some distance away from A – *without measuring it* and without having any forces acting between A and B. It is obtained solely from a measurement on A. That measurement has turned the blurred B spin some distance away into a sharp value. The EPR paper refused to accept this. How can that be?

(We remark parenthetically that this prediction of the spin direction of B can be verified by measuring it with the detector on the left. It has been fully confirmed, as will be discussed in the following section.)

Before dealing with this question, let us suppose we had chosen a different direction for making our measurement (call it the y-direction). Then the spin of B would have been changed to a sharp value along *that* direction as a consequence of measuring A. But how did B 'know' that the A spin was measured, that it was measured along the y-direction (rather than along the x-direction), and that B is therefore 'to get into' a sharp spin value along that direction?

Nobody is surprised to learn that Einstein and his colleagues, Podolsky and Rosen, refused to accept the possibility of a signal that moves with infinite speed from A to B. This would amount to some sort of 'action at a distance' a notion rejected with the advent of special relativity. They concluded instead that the B spin in that direction *must have been sharp all along*. But that is not what quantum mechanics tells us; it tells us that the *B* spin was blurred before the measurement. Therefore, quantum mechanics must be an incomplete theory! This is the key conclusion obtained by the three authors from their thought experiment.

But a further point was made by them. If the *B* spin in the *x*-direction has been sharp all along then the *B* spin in the *y*-direction (which is predicted from measuring the *A* spin in that direction) must also have been sharp all along. And if both are sharp at the same time, the uncertainty principle is violated because the spins in two different directions are observables that are *incompatible*.

These are devastating arguments. Indeed, when they were first made they caused tremendous excitement. The main point was the conclusion that the spin 'had been sharp all the time'. Einstein, Podolsky, and Rosen concluded this from the predictability of a sharp value of the *B* spin; it implied for them that the *B* spin must have *an element of physical reality*. This notion is a very basic one for EPR and is contained in their statement: 'If, without in any way disturbing a system, we can predict with certainty . . . the value of a physical quantity, then there exists an element of physical reality corresponding to this physical quantity.' That statement is exactly applicable here and leads to their seemingly inevitable conclusion.

This thought experiment was meant to convince people of Einstein's view. He appreciated the success and effectiveness of quantum mechanics but he could not accept it as a fundamental theory because of its incomplete description of reality.

What was the response to the EPR argument? It was first made by Niels Bohr and continued as a famous debate between Bohr and Einstein (Schilpp 1949). The EPR argument consists of two parts. In the first part the question is raised how electron *B* which is far away can be influenced by a measurement on electron *A*. In the second part the uncertainty principle is called into question.

Let us begin with the second part and summarize the argument: two observables which are not compatible are measured, the spin s_x along the *x*-direction and the spin s_y along the *y*-direction. Both measurements are made on the spin of *A* and the corresponding spins (up or down) of *B* are then inferred by means of the $\psi_{(whole)}$ for the entangled pair of particles. The inferred spins both have sharp values, and it is concluded that they must have been sharp all along. Since, therefore, two incompatible observables have *both simultaneously* sharp values, the uncertainty principle seems to be violated.

This argument depends crucially on the *simultaneous* presence of the sharp values of the spin observables s_x and s_y at *B*. This simultaneous presence is based on the EPR claim that *these sharp values were there all along*. Bohr's interpretation of quantum mechanics disagrees with that; it claims that they become sharp only *as a result of the measurement*. When s_x of A is measured one infers a sharp value for s_x of B; when s_y of A is measured one infers a sharp value for s_y of B. But that second measurement (of s_y) *destroys* the sharpness of s_x. Therefore they cannot be sharp simultaneously. The EPR argument against the uncertainty relation therefore fails.

Bohr's reply to the EPR argument thus points out that each measurement involves a system–apparatus interaction so that each sharp result depends on the corresponding experimental arrangement, i.e. on whether the apparatus is set up to measure s_x or s_y. Since these observables are not compatible, the second measurement destroys the sharp values which the first one produced. One cannot accept as meaningful the result of an earlier measurement; its effects on the system have been erased by the later one. Nor can one accept a future measurement that has not yet been carried out. This is where EPR went wrong.

Why does it not require a signal from A to B in order to tell B that the entanglement was broken at A? There are two answers to that question. A signal involves a transmission of energy and no energy is being transferred. If *that* had to be necessary then, indeed, a signal would be involved that could not exceed the speed of light. But what took place was only a change of state at the *same* energy: there is no violation of causality.

But more importantly, the *whole* system is involved in the interaction with the apparatus, and the whole system is changed by it, *even though it looks as if only particle A is involved*. Because of the entanglement *both* particles are affected by the measurement at the same time. This is a consequence of the holistic nature of the entangled state. The quantum mechanical answer to the first part of the EPR argument is therefore that no signal is involved because no signal is necessary.

Perhaps the most interesting conclusion is that the entanglement which is a quantum mechanical phenomenon is not restricted to the microworld. It survives macroscopic separation of the two particles! Certain quantum mechanical phenomena are observable in the macroworld. It would be incorrect to limit the domain of applicability of quantum mechanics to the microworld. In Section 11g we shall encounter several *macroscopic* pheno-mena that are not explainable by classical physics; they are effects of the quantum world.

e. *Quantum reality*

When scientists are confronted with a dilemma concerning a scientific theory they turn to experiments. Can an experimental decision be made between

Bohr's view of quantum mechanics and the very serious criticism of it as expressed by the EPR argument?

The EPR argument, along with Schroedinger's view, is based on the notions of the real world as we know it from everyday life, and as it has been strongly endorsed by the scientific progress and success of the last several hundred years. Real objects have properties with definite, sharp values. This is a fact that does not depend on how accurately our ingenuity and technology permit us to measure things (to which there is no limit in principle). Nor does it depend on whether we measure them at all. Objective reality is there whether we observe it or not. It does not depend on our measurements or even on our very existence.

This is the view of *classical realism*. Can one construct a theory for the quantum world in this spirit where all properties of particles are sharp at all times, where all objects can always be described completely, and where everything is deterministic? To this end more observables would be needed than quantum mechanics allows. Such extra observables are known as 'hidden variables' because they are not apparent, at least to quantum mechanics. We have already mentioned them in Section 10c and Note 10.8. With their aid as an addition to the observables already used in quantum mechanics, David Bohm in 1952 first constructed such a classical theory for the quantum world. In this way he thought to supply the missing elements needed to complete quantum mechanics since it was judged incomplete by Einstein and specifically by the EPR argument. In such a theory there is no entanglement and there is no holism. At worst, when those hidden variables are not observed, there is an *apparently* holistic description. That comes about when one averages over some or all of those hidden variables.

Since Bohm's suggestion, various hidden variables theories have been proposed but the matter remained moot in the absence of experimental evidence for or against them. An experimental decision between the predictions of hidden variables theories and quantum mechanics has become possible only relatively recently. Such experiments are based on an ingenious suggestion made first by the British physicist John Bell who works at the high energy particle physics laboratory CERN (the European Center for Nuclear Research) near Geneva, Switzerland. His suggestion was to make a suitable series of measurements on exactly the kind of set-up EPR used in their thought experiment as modified by Bohm. He showed how the results can be used to distinguish between the predictions of quantum mechanics and those of *any* theory of the hidden variables type as long as that theory does not involve forces that act at a distance ('locality' assumption).

This experiment involves finding the degree to which the spins of the two electrons *A* and *B* are entangled (quantum correlated). It requires a sufficiently large number of measurements of their spins (up or down) along two different directions for each electron so that the entanglement can be obtained reliably from suitable averages. The experimental result for the degree of entanglement can then be expressed in terms of a single number S, and this number can be compared with the predictions by both quantum mechanics and hidden variables theories.

John Bell proved that this number S cannot exceed 2 according to any of these 'local' hidden variables theories while the quantum mechanical prediction *can* exceed 2 if only the angles between the directions in which the spins are measured (our x and y directions) are suitably chosen.

It is most gratifying that such a crucial but very difficult experiment can actually be performed. And it is not surprising that various variations of that experiment have been tried since Bell's idea became known. The most accurate and therefore the most convincing of these was concluded only a few years ago in Paris [11.6]. For the particular choice of angles made, the quantum mechanical prediction is $S = 2.70$ while the observed value is $S = 2.697 \pm 0.015$. One sees the excellent agreement with quantum mechanics. This observed value is (well outside experimental error) significantly different from the predictions of the local hidden variables theories, $S \leq 2$.

The Paris experiment was later repeated in such a way that there was not enough time for a signal from particle A to reach particle B even if that signal were going with the speed of light. Particle A could not 'inform' B that its spin has been measured, along which particular direction it has been measured, and what the result has been. This second experiment also fully confirmed the quantum mechanical prediction and contradicted beyond any experimental error the hidden variables theories.

Together with the results of earlier experiments, the above results provide very strong empirical evidence against local hidden variables theories. It still leaves 'non-local' theories of this kind as a logical possibility but, as already mentioned in Section 11c, very few scientists indeed still believe them today since no experimental support exists in their favour.

The above empirical evidence also suggests very strongly that the EPR conclusion concerning the incompleteness of quantum mechanics is not tenable. Although Einstein, Schroedinger, and others were on familiar grounds with their classical reality concepts that played such a crucial role in the EPR argument, that argument has now run afoul of experimental evidence. One must conclude that their notion of reality, classical reality, cannot be maintained in the quantum world.

What, then, replaces the classical notion of reality in the quantum world?

The interpretation of quantum mechanics that resulted in the 1920s, the

Copenhagen interpretation (see Section 4a) provided some answers to this question. Unfortunately, that interpretation is not a clearly stated view but refers to a collection of statements made at various times by different people which do not fully agree with one another. It is however fair to say that the Copenhagen interpretation falls into a range of points of view which lies between instrumentalism and realism but which is closer to instrumentalism.

According to the instrumentalist view, as has already been discussed in Section 4a, all that theory can give us is knowledge about a possible logical relationship between the phenomena we observe. It says nothing about what there actually is in nature, and it even doubts that such a question belongs in science. It is an extreme epistemic view of science leaving the ontology of physical theories largely to the philosophers. For instance, the electron is considered to be only a theoretical entity inferred from observations. It does not command a right to be called 'real'; it is only an abstract mathematical construct and does not necessarily refer to anything real. Niels Bohr at one time is supposed to have expressed this view most bluntly: the quantum world need not exist.

In the instrumentalist view, there is little room for reality. Only the immediate sense experiences are real and everything else is inferred and depends on the theory one happens to choose for its 'explanation'. In this view reality with its common-sense meaning in the classical world is not replaced by anything in quantum mechanics; it is eliminated. Observables are blurred and particles are entangled because that is what our best theory deduces from our observations. Whether they actually *are* blurred or entangled is a philosophical question and lies outside science. The electron that gives an interference pattern in the double-slit experiment (Section 10b) does not exist until one sees a blip on the fluorescent screen: reality is created by the observer! This is an absurd conclusion indeed.

Such an extreme empirical and positivist view makes it of course very easy to accept the strange new features of holistic systems and of the quantum world in general. For Schroedinger that was too easy a way out. He called it a 'dictatorial help' and 'the supreme protector of all empiricism'. However, although he struggled very hard to find an interpretation of quantum mechanics consistent with his classical realist views, he did not succeed. Nor did Einstein.

It thus appears that if there is an interpretation of quantum mechanics in a realistic sense, that notion of reality would have to be quite different from the classical notion of reality. Let us therefore call this new and different reality by a special name: *quantum reality*.

Quantum realism rejects the instrumentalist view. It claims that electrons do exist. Their being is in no way different from the being of macroscopic objects. Such objects also do not always have sharp attributes: a pair of tumbling dice have no sharp orientation and have a practically unpredictable motion. To be sure, if one goes to enough trouble, a careful study will show that they do have a sharp orientation and that their motion can be predicted. But is this really necessary for the reality of their existence?

This brings us to the heart of the matter: there are physical properties that are not essential for the identification of an object. We do not identify dice as real objects on the basis of their having a definite position; similarly, we do not identify a quantum particle as a real object on that basis. A single free electron is real whether or not it happens to have a sharp position or a sharp speed. We identify it by its mass, its electric charge, and its magnitude of spin. And when we ask what quantum mechanics has to say about those properties, we find that *those particular observables are never blurred but always have sharp values* [11.7]. Electrons are really there according to the quantum realist view, even though many of their properties (position, speed, etc.) may be blurred, and even though their motion is predicted in a probabilistic fashion. Similar statements can of course also be made about other quantum systems.

In the same vein, the holistic properties of quantum systems and their entanglement are considered to be actually existing features and not only an expression of our (possibly incomplete) knowledge of them. There is an ontology to quantum mechanics that is denied in the instrumentalist interpretation: *the entangled (blurred) state really exists*.

The conceptual changes necessary for the transition from classical reality to quantum reality are not easy to make. It requires a broadening of the notion of classical reality. A real object is no longer restricted to sharply defined attributes only. The identification of reality with the sharpness of an object's properties must be left behind. And the probabilistic nature of the world must be accepted where 'determinism in principle' no longer holds. It is replaced by the weaker determinism of probability amplitudes.

In summary, there are two essential differences between quantum systems and classical ones: quantum systems have some (but not all) of their properties blurred while the properties of classical systems have all sharp values; and quantum systems develop in time in a probabilistic way while classical ones are deterministic (with certain important exceptions hinted at in Section 10c.) But both can claim real existence. The quantum world is as real as the classical world.

In fact, there is only one world and only one reality: what we call the quantum world is simply the collection of those phenomena that obey the laws of quantum mechanics. The classical world, by contrast, comprises those in which the quantum nature of the elementary phenomena is so submerged, so hidden, that we can see only an effective behavior that is quite different from quantum mechanics: it is the classical behavior. There exists of course a *continuous* transition as one goes from one to the other although such intermediate cases are often very difficult to analyze.

It is exactly at this point where the instrumentalist view fails: if the quantum systems have no claim to reality but only classical ones do, where is the borderline between the two? Is the quantum world only the fiction of the imagination of theoretical physicists? Is reality created by the observer?

It is quite unimportant to what extent the Copenhagen interpretation is an instrumentalist view. What does matter is that while most working scientists claim to accept the Copenhagen interpretation, there is little doubt that most of them intuitively work with a quantum realist view: they investigate the real world.

f. *Quantum logic*

The revolutionary nature of the new concepts forced upon us by quantum mechanics has given rise to a considerable amount of discomfort. The theory has been called strange, weird, and even incomprehensible. Since it violates common sense and offends our sensibilities, some people have even asserted that 'nobody understands quantum mechanics'.

Now it is important that the interpretation of quantum mechanics became possible only *after* the formal mathematical structure had been successfully applied to a large variety of problems. There were mathematical symbols which had a well-defined meaning within the framework of pure mathematics but which also demanded a *physical* interpretation. Foremost among them was the symbol ψ. The mathematical structure of the theory thus directed the search for the understanding of its physical meaning.

New light was shed on this problem about ten years after the basic theory got into place. Two mathematicians, Garrett Birkhoff and John von Neumann, discovered a strange and unexpected similarity between the mathematics of quantum mechanics and the mathematics underlying logic.

That our reasoning and argumentation is based on certain 'laws of thought' was already quite apparent to Aristotle (384–322 BC). These laws govern the way we infer a statement from other statements. They have been studied by many people over the years and the discipline became known as *logic*. But not until the middle of the last century was it discovered that this logic is expressible as a kind of algebra. This was done by George Boole (1815–64). He labeled different statements ('propositions' in the language of logic) by letters such as A, B, C, etc., and then he introduced *logical symbols* for such concepts as 'not', 'and', and 'or' (Table 11.1) which function in a similar way to the symbols for 'minus', 'plus', and 'times' that are used in ordinary algebra (as well as in arithmetic). That newly constructed algebra, *Boolean algebra*, is clearly not the same as ordinary algebra. But it does represent Aristotelean logic faithfully.

Birkhoff and von Neumann of course knew Boolean algebra. And they found that the mathematics of quantum mechanics has an algebraic structure that is similar to Boolean algebra but is not the same. It represents a logic which is *different* from Aristotelean logic; it is one of many possible *non-Aristotelean logics*. This new logic was called *quantum logic* [11.8].

What does this mean? One way of putting it is to say that 'if we were thinking according to quantum logic rather than according to Aristotelean logic, quantum mechanics would not seem strange to us'. Unfortunately, this does not help very much because we *do* think according to Aristotelean logic. In fact, when we talk about non-Aristotelean logic we do it by using our usual Aristotelean logic. However, scientists who have been using quantum mechanics a great deal have become so well acquainted with it that in a sense they do think in this fashion when they do quantum mechanics.

How does one formulate logical statements (propositions) in quantum mechanics? Propositions are to be either true or false. If one has more complicated statements, one has to break them down into those that have only true–false answers. In quantum mechanics such answers require *measurements*. Thus when *A* is the proposition 'the spin of the electron is up' a measurement determines whether or not this is a true statement. Quantum logic therefore deals only with epistemic statements, statements that are found true or false upon measurement.

In order to get the flavor of this new logic it is worthwhile to compare quantum logic and ordinary (Aristotelean) logic. They have in common that a statement is either true or false. There is no third choice; this fact is

Table 11.1. *Symbols for logic.*

statement	A	letters designate statements; for example, A stands for 'Schroedinger's cat is dead'
'not'	−	this symbol negates a statement; for example, $\overline{A}$ stands for 'Schroedinger's cat is not dead'
'and'	∧	combines two statements so that $A \wedge B$ is true if and only if A and B are both true
'or'	∨	combines two statements so that $A \vee B$ is true if and only if at least one of the two statements is true

sometimes called *the law of the excluded middle*. Quantum logic as well as ordinary logic is *two-valued*. One can express this symbolically as follows. Using the symbols of Table 11.1,

$$A \vee \bar{A} = 1.$$

This equation states: the proposition that either '*A* is true' or '*A* is not true' is always true; the 1 indicates 'always true'.

But although quantum logic is two-valued, and does satisfy the above equation, it is also, *in a certain weak sense, not* two-valued: a third possibility exists. We shall see this shortly.

One of the fundamental laws of our conventional logic is the *distributive law*. It is formally very similar to the distributive law of ordinary algebra where we have

$$A \times (B + C) = (A \times B) + (A \times C).$$

In Boolean algebra (Aristotelean logic) we have the analogous formula

$$A \wedge (B \vee C) = (A \wedge B) \vee (A \wedge C).$$

To see the meaning of this law of logic we can substitute statements for *A*, *B*, and *C*. For example:

Let *A* stand for 'in the double-slit experiment (Fig. 10.1) the electron hits the screen at point *P*'.

Let *B* stand for 'the electron has gone through the upper slit'.

Let *C* stand for 'the electron has gone through the lower slit'.

Using the meanings of the logical symbols as given in Table 11.1 the algebraic expression *A* (*B C*) then means:

The electron hits the screen at point *P and* the electron has gone through at least one of the two slits.   (First Proposition)

On the other hand, the algebraic expression (*A* ∧ *B*) ∨ (*A* ∧ *C*) means:

At least one of the two propositions is true: 'the electron hits the screen at point *P* and has gone through the upper slit' or 'the electron hits the screen at point *P* and has gone through the lower slit'.

(Second Proposition)

Now we must remember that all these propositions are true or false on the basis of measurements. Therefore, while the first proposition does not involve a measurement to find out through which slit the electron has gone, the second proposition does!

According to conventional logic, the two propositions (given symbolically by the two algebraic expressions) are equivalent; they mean the same thing and the distributive law holds. And that is exactly what we would expect if this were a classical experiment (with tennis balls rather than with electrons): when the experiment is repeated many times and *P* ranges over the whole baffle the result is Fig. 10.2: one can measure through which slit each one of the balls passes without in any way affecting the distribution of hits on the screen.

But in quantum mechanics (experiment with electrons) when both slits are open *and one does not measure through which slit the electron goes*, one obtains something quite different, namely the interference pattern of Fig. 10.3(*a*). This corresponds now to the first proposition. It is not the same as the experiment in which one does measures through which slit the electron goes (required for the second proposition). Therefore, the two statements are not equivalent in quantum mechanics and the distributive law of logic does not hold.

This example demonstrates how quantum logic differs from conventional (Aristotelean) logic, and how it accounts correctly for the strange relations between quantum phenomena.

Let us now consider the special case in which the above proposition C is the proposition $\overline{B}$ (meaning not B). The distributive law then reads

$$A \wedge (B \vee \overline{B}) = (A \wedge B) \vee (A \wedge \overline{B}).$$

Because of the law of the excluded middle, the left side of this equation is $A \wedge 1$ which is just A. The right side states: either 'A and B together are true' or 'A and "not B" together are true'. In classical logic this is the same as 'A is true' so that the distributive law holds. In quantum logic this is not the case: the right side of the equation is not the same as 'A is true'. In this (weaker) sense the law of the excluded middle does not hold in quantum mechanics.

If this argument seems rather abstract, the reader should use the above propositions for A and B and verify that the above distributive law indeed does not hold in quantum mechanics.

Traditionally, it has always been believed that logic is given *a priori*; it is not empirically determined. Yet, here one seems to have a logic which is determined from the experiments on the quantum world. It was not known before. Is quantum logic empirical?

This question seems to have some similarity to the question whether geometry is empirical in view of the use of non-Euclidean geometry in Einstein's gravitation theory (Chapter 7). What was at one time considered 'obvious', unique, and beyond question (Aristotelean logic, Euclidean geometry) was found to be only one of several different alternatives (non-Aristotelean logics, non-Euclidean geometries). What actually occurs in nature is not the 'obvious' (the one familiar from everyday experience) but one of the unexpected varieties of logic and geometry. It would seem that the question 'which logic and which geometry actually do occur in nature' has become an experimental question.

However, this parallel is defective. The non-Euclidean geometry of space–time is fundamental to gravitation theory. Einstein's theory is inconceivable without it; the theory is built on it. Quantum mechanics is not built on quantum logic. That logic is an afterthought. The theory can be carried through and comparisons with observations can be made without ever referring to quantum logic. One can even hold that quantum logic is just an amusing equivalent to the mathematics of quantum mechanics without having any bearing on the theory. No such statement can be made about the geometry of space–time.

The empirical basis of the geometry of space was evident already to Carl Friedrich Gauss, the great mathematician who was Riemann's teacher and who first knew about non-Euclidean geometry. He attempted an experiment to determine whether space is Euclidean by measuring the angles of a large triangle formed by three light rays connecting three mountain tops. His idea was in principle correct, but the accuracy of his experiment was far insufficient to detect the small deviations from the Euclidean sum of 180 degrees.

Thus, while geometry is essential to the theory of gravitation, quantum logic is not essential to quantum mechanics. And beyond that, quantum logic deals only with the epistemology of the quantum world because it has to do only with empirical statements. It has nothing to say about its ontology and is completely neutral to whether one adopts an instrumentalist or a quantum realist view. It thus adds little to a philosophical understanding of the subject.

Nevertheless, one can learn a great deal from quantum logic. It adds a very interesting and novel aspect to quantum mechanics; it provides an equivalent to the mathematical structure of the theory (though insufficient for computations); and it shows how the 'weirdness' of quantum mechanics emerges from its mathematics. But it does not contribute to the deeper problem of the ontology of the quantum world.

g. *Macroscopic quantum phenomena*

The quantum world is often identified with the world of the very small. Only the atomic world is thought to be governed by quantum phenomena. This belief is erroneous. Quantum phenomena *can* occur on a macroscopic scale. Various discoveries in our present century have proven this. In the following we shall consider some of these.

Perhaps the first thing that comes to mind is the double-slit thought experiment which played such an important role in analyzing the difference in behavior between classical and quantum particles (Sections 10a and 10b). An actual experiment of this type involving macroscopic distances has recently been carried out. It was done with neutrons rather than electrons.

Neutrons are electrically neutral particles that are about 2000 times heavier than electrons. They occur as constituents of atomic nuclei (Note 6.15 and Section 12b). The momenta of the incident neutrons were such that their de Broglie wave lengths were about 2×10^{-9} m. An interference pattern was observed that is qualitatively the same as would be obtained by corresponding light waves (electromagnetic radiation). It agrees exactly with what is predicted by quantum mechanics (Fig. 10.3(a)).

Some macroscopic quantum phenomena occur at very low temperatures. There is a law in the theory of heat called the third law of thermodynamics. It states that there is a lowest temperature that can never be reached although one can get very close to it; any physical system at that temperature would be in a state of maximum order [11.9]. That temperature, the *absolute zero*, is −273 degrees centigrade. It is often convenient to measure temperature from absolute zero on up, and one calls it 'degrees Kelvin' (after the British physicist William Thomson, Baron of Kelvin (1824–1907)). Thus, at normal pressure, ice melts at +273 degrees Kelvin and water boils at 373 degrees Kelvin. Macroscopic quantum phenomena occur at a few degrees Kelvin. The two phenomena of interest to us are superconductivity and superfluidity.

Superconductivity is a very baffling phenomenon. When a metal such as a wire conducts an electric current, it heats up. One can easily understand this: the electrons in the metal whose motion constitutes the electric current (the 'conduction electrons') are not bound to atoms but can 'float' in the metal. When they are made to move by some applied voltage they encounter resistance because they collide with the atoms of the metal. This causes the electrons to lose energy which is taken up by the atoms. As a result, the atoms increase their speed and this increase of speed appears macroscopically as an increase in the temperature of the metal. Therefore, any wire through which a current flows heats up. But when certain metals are cooled down to just a few degrees Kelvin, their behavior changes dramatically: they become superconductors in the sense that they conduct electricity *without* any increase in temperature whatever.

Superconductivity happens below a certain temperature (characteristic for each metal) called the 'transition temperature'. At those temperatures there is no loss of energy by the conduction electrons. Once an electric current is set up (for example, in a ring-shaped superconductor), it will continue to flow without the aid of a battery, and it will not diminish in intensity. In some laboratories currents have been observed to flow for many years without detectable weakening.

This phenomenon is quantum mechanical in nature. It cannot be fully understood in terms of classical physics. In very general terms the explanation is as follows. The conduction electrons in the metal become bound to one another in pairs. The force which causes this binding is mediated by the atoms of the metal. They attract the electrons in such a way that there is an effective binding force between pairs of them. The electrons in each pair move in opposite directions and have their spins aligned oppositely so that the pair as a whole has total spin angular momentum zero. The quantum-mechanical probability amplitude ψ makes each electron pair stick together (cohere) over relatively very large distances thousands of atoms apart. Different pairs can therefore overlap considerably.

This pairing of the conduction electrons below the characteristic temperature establishes a certain amount of order among them which is absent above that temperature. And this order is a *long-range order*; each pair has a very extended probability amplitude ψ. When an electron loses a small amount of momentum, its pair partner will experience a corresponding increase in momentum because of the coherence of the pair. The pair as a whole will not lose momentum and will continue moving undisturbed. It takes a certain minimum amount of disturbance for pairs to break up and lose energy.

But beyond that, the many pairs cooperate with one another; they are to a very large extent locked into one another's motion. This locking-in is also a characteristically quantum mechanical feature. It has to do with each pair acting like a single particle of angular momentum zero. Such an ensemble of particles tends to be in a state where almost all particles have the same velocity. When a battery imposes a voltage on the superconductor, all pairs will move together in a locked-in way. And this lock-in occurs over *macroscopic* distances. As a result, it is no longer possible for individual electrons to lose energy; they are being forced to move along with all the others. No energy is transferred to the atoms, the metal does not heat up, and the electrons continue moving even after the voltage ceases without experiencing any resistance: one has superconductivity.

When the temperature is raised on a superconductor, the atoms are forced to move faster and more randomly, eventually breaking the pair bonds of the electrons and destroying the superconductivity.

It is perhaps not without interest to note that superconductivity had been known for almost 50 years until it could be explained. Not that there was a dearth of suggested theories. They came at a rate of a few almost every year. But they all failed because the intrinsically quantum-mechanical nature of this effect and the surprising pairing of the electrons was not realized for a very long time. There are many interesting phenomena related to superconductivity that such a theory has to explain too, but we cannot discuss these here.

Applications of the property of superconductivity includes the manufacture of very strong electromagnets and the transmission of electricity without heat loss (but at the cost of having to cool the wires).

Another surprising phenomenon is *superfluidity*. At normal pressures the well-known gas helium turns into a liquid when the temperature is lowered sufficiently, just like all other gases. But unlike all other liquids, liquid helium does not become solid no matter how closely one approaches absolute zero (zero degrees Kelvin). This is related to the fact that liquid helium (called He I), which is like any other liquid down to about 2.2 degrees Kelvin, changes into a very different kind of liquid (called He II) below that temperature. This is a change of *phase* like the change from a liquid to a solid or from a gas to a liquid because of the complete change in properties. The transition temperature of 2.2 degrees Kelvin is known as the 'lambda point' [11.10].

The properties of the liquid called He II are very surprising indeed. For example, it is not at all viscous like other liquids so that it can leak out of a porous porcelain jar which could hold He I without any problem. It also has a fantastically high heat conductivity about a million times higher than He I and higher than metals such as copper. One can observe this by watching He I boil above the lambda point and suddenly stop boiling as the temperature is reduced below that. The gas bubbles form as a result of heat concentrations in different places. Such concentrations cannot occur when heat is conducted away very rapidly due to the very high heat conductivity in He II so that it has at all times a uniform temperature throughout. And there are other very curious properties of He II.

What is the explanation of this strange phase of helium? The answer lies again in quantum mechanics. As it is cooled below 2.2 degrees, He I suddenly changes into a mixture of two fluids, a normal fluid and a superfluid. The normal fluid is like He I but the superfluid is very different and is responsible for the above strange properties. This superfluid is the lowest energy state of all its helium atoms together, one probability amplitude for the whole macroscopic substance. Again, as in superconductivity, this is the result of the quantum behavior of an ensemble of particles of zero angular momentum (the helium atoms have no spin). They all tend into the same quantum state and attract one another even down to absolute zero. Consequently, they move together without internal friction (no viscosity) and show superfluid behavior. The normal fluid can be regarded as a higher energy state of the superfluid. As the temperature is raised an increasing fraction of He II is in the normal fluid state until it becomes 100 per cent at the lambda point.

Because of the dominant importance of quantum mechanics for He II that fluid is called a *quantum fluid*. Its theory is still not completed today. Certain features (not discussed above) can still not be accounted for quantitatively. But the above model is at least a very good approximation since it explains most of the experimental results correctly.

Annotated reading list for Chapter 11

Bohr, N. 1963. See annotated reading list for Chapter 10.

d'Espagnat, B. 1979. 'The quantum theory and reality.' *Scientific American* Nov.: **158**. A very good review of experiments of the EPR–Bohm type before the Paris experiment.

Heisenberg, W. 1958. *Physics and Philosophy*. New York: Harper. Here are the views on quantum theory by one of the founders of the subject.

Jammer, M. 1974. *The Philosophy of Quantum Mechanics*. New York: John Wiley. A valuable reference to the various interpretations of quantum mechanics that have been suggested over the years.

Mendelsohn, K. 1977. *The Quest for Absolute Zero*. London: Taylor and Francis. The history of the scientific efforts to explore and understand low temperature physics is related by one of its leading participants. This subject relates to section 11g.

Mirman, N. D. 1985. 'Is the moon there when nobody looks? Reality and quantum theory'. *Physics Today* April: pp. **38–47**. A simple model experiment is constructed which demonstrates beautifully how counter-intuitive the results of the EPR–Bohm experiment really are.

Pagels, H. R. 1982. See the annotated reading list for Chapter 10.

Rohrlich, F. 1983. 'Facing quantum-mechanical reality'. *Science* **221**: 1251–5. A non-technical review of Bell's theorem, its test by the Paris experiment, and a summary of some of the basic tenets of quantum mechanics.

Schilpp, P. A. 1949. *Albert Einstein: Philosopher-Scientist*. New York: Tudor Publishing Company. A large collection of descriptive and critical essays on Einstein's work and Einstein's reply. Of special interest is the Bohr–Einstein debate on the interpretation of quantum mechanics.

12

The present state of the art

a. *The marriage between quantum mechanics and special relativity*

Newtonian mechanics, despite its spectacular successes, was found to be wanting: it is not applicable to very fast motion (motion close to the speed of light). We have seen how the mechanics of special relativity theory superseded Newtonian mechanics.

A similar fate awaited quantum mechanics; it, too, had spectacular successes (examples were seen in Section 11g). But quantum mechanics is also not applicable to very fast moving quantum particles. Indeed, it is the quantum version of Newtonian mechanics: it describes quantum particles with respect to inertial reference frames that are related by *Galilean* transformations. The laws of quantum mechanics are invariant only under *those* transformations; they are not invariant under the Poincaré transformations of special relativity (Section 6g). That means that they hold only for motions slow compared to the speed of light. For very fast particles we need the quantum version of special relativistic mechanics.

That new theory can be regarded as a generalization of quantum mechanics. Or it can be regarded as a superseding theory that unifies quantum mechanics with special relativity theory, very much in the spirit of the unification of theories in Section 4b. Such a unification would therefore be a higher level theory. The last half-century of research in fundamental particle physics was in fact dominated by the problem of finding such a theory. It aimed at the 'marriage' of the two pillars of modern physical science, quantum mechanics and special relativity.

In order to understand this unification on the conceptual level we recall the two dualisms that were removed by these two theories. In early quantum mechanics the particle–wave dualism played a dominant role as an obstacle to a deeper understanding. The removal of that dualism came about by abandoning both of these classical concepts in favor of a more sophisticated notion, the quantum particle, which object is neither a particle nor a wave but can behave as either one of them under suitable conditions.

The other dualism dominated the physical sciences preceding special relativity. A sharp distinction was made between matter and energy. The theory of special relativity taught us that no such sharp distinction exists. Matter, which by definition has mass, consists of mass energy and can be converted into other forms of energy (and *vice versa*). An object of mass m at rest has energy $E = mc^2$.

A unification of quantum mechanics and relativity must involve also a unification of these two dualisms and their removal. On one side of these dualisms we have 'particle' and 'matter', which are localized entities. On the other we have 'wave' and 'energy', entities which are in general not localized. The latter entities have been encountered before in the notion of 'field' [12.1]. For example, electromagnetic waves are oscillating (vibrating) electromagnetic fields; they also carry energy (and momentum). The unification of the two dualisms which preceded quantum mechanics and special relativity thus leads to the new dualism of matter–particles (localized) on one side and wave–fields (not localized) on the other (Fig. 12.1).

The new unified theory resolves this new dualism: the new concept of *quantum field* is neither a matter–particle nor a wave-field (both understood in the classical sense). But although it is neither, it can behave as either one of them under suitable conditions.

The unified theory is called *quantum field theory* (Fig. 12.2). It is a relativistic theory in the sense that two inertial reference frames that move with constant velocity relative to one another will observe the same laws of motion and are related by a Poincaré transformation (Section 6g). Quantum fields, however, differ from fields of classical physical sciences such as the

Fig. 12.1. The removal of the mass–energy and the particle–wave dualisms and their unification into a single particle–field dualism. The latter is removed in quantum field theory.

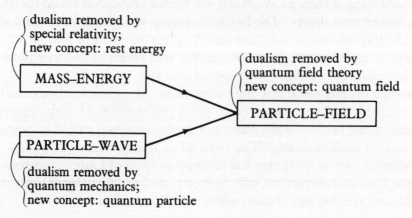

classical electromagnetic fields or the velocity field of a fluid [12.1]. Each quantum field is associated with a particular type of quantum particle. It resembles in this respect a probability amplitude ψ of quantum mechanics. Its square gives information about the probability for finding the particle at a certain place at a certain instant of time.

It is important to realize that this somewhat vague conceptual picture of the unification of quantum mechanics and special relativistic mechanics emerged only after the appropriate mathematical structure of the unified theory, namely of quantum field theory, was understood. A unification on the conceptual level could not possibly have been carried out before that; it needed mathematics as a guide.

In quantum field theories the forces between the quantum particles are expressed by interactions between the quantum fields. This interaction results in much more difficult mathematics than is contained in either quantum mechanics or special relativity theory. For example, many fewer problems can be solved in quantum field theory without the aid of computers. But those problems that have been solved have shown agreement with experiments beyond all expectations that scientists had even as recently as half a century ago [12.2]. The particular quantum field theory that shows this remarkable success is the one that deals with the electromagnetic forces. It is called *quantum electrodynamics* and it is the quantum version of Maxwell's classical electrodynamics applied to quantum particles that can move very fast, even very close to the speed of light.

Quantum electrodynamics started with a work by the British physicist Paul Adrien Maurice Dirac (1902–84) who later occupied the Lucasian Chair at Cambridge University, the position that had been held much earlier by Isaac Newton. Dirac was able to generalize the Schroedinger equation of quantum mechanics so that it became applicable also to fast elec-

Fig. 12.2. How special relativity and quantum mechanics supersede Newtonian mechanics, and how they are unified into a single theory.

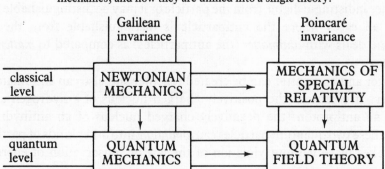

trons (1928). This new equation is now known as the *Dirac equation*. It describes the causal development in time of the electron (or rather of its potentialities) just like the Schroedinger equation does in quantum mechanics. But the electron is now described by a quantum field rather than by a probability amplitude ψ.

This quantum field, the electron field, can interact with the electromagnetic quantum field. The theory that deals with this interaction is quantum electrodynamics. It permits electrons (as quantum particles) to exert forces on one another by means of electromagnetic forces (in their quantum form). The electromagnetic field thus *mediates* these forces as in the classical Maxwell theory. But in quantum electrodynamics it is all done on the quantum level.

The Dirac equation led to a number of most remarkable consequences. One was that it had 'built in' the spin property of the electron. The Schroedinger equation of quantum mechanics originally did not describe the electron as a particle with a spin. When it later became known from experiment that it does have a spin, this property had to be added to the Schroedinger description of the electron 'by hand'. No such *ad hoc* device is necessary when the Dirac equation is used.

Another consequence of this equation was that for every solution there is a second solution which describes the same electron (same mass, spin, etc.) but with its electric charge of opposite sign. At that time only electrons with negative electric charge were known. But the Dirac equation predicted also the existence of electrons with positive electric charge. To the great triumph of the Dirac equation, positively charged electrons were indeed discovered only four years later in the cosmic radiation [12.3].

These *positrons* as the positively charged electrons are called were the first example of *antiparticles*. As was soon realized, all relativistic quantum particles come in pairs like that: for each particle there is an antiparticle. If the particle is electrically charged the antiparticle has the opposite electric charge. If the particle is not electrically charged then the antiparticle may be either indistinguishable from the particle or it may be distinguishable from it. In all cases where the antiparticle is distinguishable from the particle one deals with *antimatter* (the antiparticles) as compared to *matter* (the particles).

Matter and antimatter can be created together and also can annihilate one another. If a proton (the positively charged nucleus of a hydrogen atom) meets an antiproton (the negatively charged nucleus of an antihydrogen atom) these two quantum particles can *annihilate* into other kinds of particles. This takes place without violation of the law of conservation of energy: a proton of mass m at rest has an energy mc^2 according to the special theory of relativity (Section 6e), and so does an antiproton because it has exactly the same mass. Consequently, their annihilation at rest produces particles of total energy $2mc^2$. This has been checked and confirmed repeatedly by experiments.

The particle–antiparticle annihilation process can also be reversed: two photons of at least $2mc^2$ energy can collide, annihilate, and produce a proton and an antiproton. A particle–antiparticle pair can be *created*. If the two initial photons have more energy than the minimum $2mc^2$ then the proton–antiproton pair will be created not at rest but with a certain amount of energy of motion (kinetic energy). They will fly apart. Pair annihilation and pair creation has been observed under suitable conditions for many different types of particles.

The mathematics of quantum field theory has here led us to an entirely new and unexpected concept: the notion of antimatter. And that notion was later found to describe reality correctly: the predicted phenomena were indeed observed. Both the qualitative and the quantitative predictions of quantum field theory were confirmed.

Today quantum field theory is the backbone of our understanding of fundamental particles and the forces between them. The great success of its application to electromagnetic forces suggested a similar application to the weak forces and to the strong forces (Section 4b). The application to the *weak* forces led to the unification of electromagnetic and weak forces into the new *electroweak* theory (Fig. 4.3). But neither the experiments nor this theory permit a comparison to the same fantastic accuracy that was found in the case of quantum electrodynamics.

The application of quantum field theory to the *strong* forces is called quantum chromodynamics (see Section 12c). It is a considerably more difficult theory than the theory of the previously studied forces. The unification of the quantum field theory of strong forces with the quantum field theory of electroweak forces is called *grand unified theory* (see Fig. 4.3). It has so far not succeeded in conformity with observations [12.4].

Finally, the problem of finding the quantum version of the gravitational forces is even farther from our present understanding. It is an extremely difficult problem because it requires the quantum version of Einstein's gravitation theory (Chapter 7). A conceptual understanding of this theory, quantum gravity theory, will have to wait until its mathematical framework is established. We expect its mathematics to lead us into the unknown just as the mathematics of quantum mechanics did earlier and the mathematics of quantum field theory did more recently.

b. *The onion of matter*

In 1982 one of the most instructive and beautiful books in years was published. It is called *The Powers of Ten* and deals with all sizes of things in the universe. The photographs are by Charles and Ray Eames, and the text is by Philip and Phylis Morrison (see Reading List at the end of this chapter for a complete reference).

This relatively thin book (about 150 pages) contains 42 color plates which show things from the largest to the smallest sizes known, from 10^{25} meters to 10^{-16} meters [12.5]. It thus covers the universe from regions of space so large that whole galaxies of stars appear only as barely visible points, to regions of space so small that the most elementary building block of matter, the quark, would just barely fit into it.

When we consider matter of the size we know from everyday life and break it into smaller and smaller components we are dealing only with a fraction of this enormous array of sizes, barely more than one-third of the powers of ten depicted in that book. The major part of it deals with much larger things than our human size: the earth, the solar system, our galaxy, and the vastness beyond that. But that smaller part from human size on down presents an enormous richness and diversity that has fascinated and occupied scientists and artists alike. We cannot grasp it all at once, and we must consider it piecemeal, digest it in small bites, as it were. A crude way to characterize these bites is to use their size in powers of ten.

The scientific enterprise is correspondingly divided into specialties according to size. For example, living organisms occupy a very wide range of sizes from enormous Sequoia trees to microorganisms more than ten million times smaller. Specialities in *biology* (which includes botany and zoology) range from vertebrae zoology to microbiology, from the most complex to the simplest organisms. Their variety is indeed stunning: among insects alone about one million species have been identified. In terms of the microscopic

Table 12.1. *The onion of matter in terms of successively smaller building blocks.*

Molecules
Atoms
Atomic nuclei
Fundamental particles
Elementary (?) particles

structural units of living matter, the *cells*, the simplest organisms, the protozoa, involve only one single cell. The diverse nature of individual cells in turn resulted in a separate branch of science called cytology.

Below this wide array lies the domain of *chemistry* ranging from the huge macromolecules such as proteins in living matter (of interest to the biochemist) down to the diatomic molecules such as nitrogen and oxygen in the air we breath. Over a million organic chemical compounds [12.6] have been identified, and many remain to be analysed. The structural units of chemistry are the *molecules*. Their size ranges over approximately four powers of ten (see Fig. 12.3 and Table 12.1).

The next layer of sizes is the domain of the *atom*. A tremendous simplification awaits us here: the many millions of chemical compounds can all be understood in terms of only about one hundred elements [12.7]. The very great diversity of chemical compounds results entirely from the type of atoms which constitute the molecule and from the way these atoms hang together. The basic structure of atoms is also much simpler than that of molecules: a positively charged nucleus (extremely small compared to the size of the atom but carrying almost all of its weight) is surrounded by electrons (which are negatively charged). The whole structure is held together only by electromagnetic forces and most importantly by the electric attraction force between the oppositely charged particles (Coulomb's force). *Atomic physics* covers approximately four powers of ten down to the size of atomic nuclei, a few times 10^{-14} m.

The complexity of these *nuclei* resulted in a separate specialty, *nuclear physics*. It came into its own with the discovery of the neutron about half a century ago (1932). The atoms of each element differ from the atoms of all the others by their nuclei. These are primarily responsible for the chemical properties of the atoms. They also determine how many electrons may be

Fig. 12.3. Levels of science according to approximate size. There are of course no sharp boundaries between these levels.

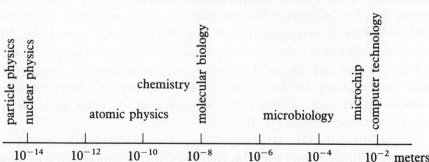

present in the atom. Thus, there are as many different nuclei as there are elements [12.8].

Having over one hundred different nuclei built up out of only two kinds of particles, protons and neutrons, is a very remarkable fact. One suspects that one may be approaching the most fundamental level of matter. There was a time when physicists thought that we might have come close to the 'center' of this *onion of matter*, these layers within layers, when there is no further deeper structure to be explored. This suspicion and pious hope received, however, a severe blow during the last quarter-century with our ability to build particle accelerators of increasingly high energy. The higher the energy of the particles accelerated in those machines the deeper they can penetrate their targets (other particles in the target atomic nuclei) and the more we can learn about their internal structure.

When two quantum particles collide, they may just change their directions and speeds as they move apart again. But if they collide with very high speed they will most often produce one or several new particles in the process. These processes can be recorded on photographic emulsions as well as by other means. As a result, this new and quickly improving technology of high-energy accelerators led to the discovery of a multitude of new particles that were produced in collisions and that had hitherto been unknown. New *fundamental particles* were thus found at an average of almost one every month. The proton and the neutron became just two of hundreds of quantum particles. It was truly an embarrassment of riches. This layer of matter that was thought to contain just the two ultimate constituents of matter, the neutron and the proton, actually contains a very great deal more. A new speciality was established, *particle physics*. It deals with objects smaller than atomic nuclei, 10^{-14} m and below.

c. *Elementary particles*

Is there no center to the onion of matter? Will new layers continue to be discovered forever? It is remarkable how much farther we have dug into the structure of matter since the ancient Greeks. Leucippos and Democritos (5th century BC) first suggested that all matter consists of smallest indivisible particles and they called them 'atomos', the Greek word for 'indivisible'. The Greek atom was found to have a nucleus, and the nucleus was found to consist of neutrons and protons, and these in turn were found to be not 'the elementary building blocks' of matter but just two of a large family of fundamental particles. Will we ever reach *elementary* particles, particles that have no further structure? [12.9].

The present state of the art emerged from a successful classification of this multitude of fundamental particles combined with a judicious use of quantum field theory applied to the large amount of experimental data collected from particle accelerator experiments. Let us summarize the present (still tentative) view on the question of elementary particles.

The force between two charged particles is described classically by a field, the electromagnetic field. Each electric charge produces such a field of force. This field spreads from the particle in all directions with the speed of light. When it reaches another charged particle it exerts a force on it. The electromagnetic field thus mediates the electric and magnetic forces from one particle to the other. For particles that are bound together like the proton and the electron in a hydrogen atom the electromagnetic field acts like an 'electromagnetic glue' covering the distance between the particles with the speed of light. In quantum electrodynamics this field can also manifest itself in the form of particles, namely the photons. They can thus be thought of as 'glue particles'. These glue particles are running back and forth between the electron and the proton and provide the force between them.

The quantum field theories of other forces (the weak and the strong forces) are fashioned in a way analogous to quantum electrodynamics. In all cases there is a field in the standard description that mediates the force between the interacting particles. And in quantum field theory this field corresponds to glue particles so that each force has its own characteristic glue particles. We can thus distinguish between the particles that are the *source* of the glue particles (such as the electrons and the protons) and the *glue particles* themselves (such as the photons). The four fundamental forces are each associated with different kinds of glue particles (Table. 12.2). While the electromagnetic and the gravitational force have only one glue particle each the weak force has three and the strong force that holds the atomic nucleus together has very many such glue particles (mesons by name).

The source particles come in two very distinct kinds light ones and heavy ones. Using the corresponding words of Greek origin, they are called, respectively, *leptons* and *baryons*. There are three light ones and many heavy

Table 12.2. *The glue particles.*

Force	Name	Number
Electromagnetic	Photon	One
Weak	Weak gluons	Three
Strong	Mesons	Many
Gravitational	graviton	One

ones in addition to the proton and the neutron (Table 12.3). The baryons and the mesons are the only particles in the two tables that relate to the strong forces. They are therefore given a common name; they are called *hadrons* after the Greek word for 'strong'.

But now we find that the hadrons (the mesons and the baryons) are the only particles that occur in large varieties. There are only a few glue particles for each force, as would be expected on the building block level except for the strong force where there are far too many mesons; and there are only a small number of leptons which could thus be considered candidates for building blocks but there are many too many baryons. One would expect only a small number of mesons and of baryons if they were truly *elementary* particles. This problem is resolved according to current thinking by the theory of the *quark* [12.10].

Neither mesons nor baryons can be considered elementary because there are far too many of them. The quark theory regards these particles as composites of quarks, the baryons consisting of three quarks and the mesons consisting of one quark and one antiquark. The quarks themselves are thought to be fundamental building blocks, elementary particles, but of a very special kind. They are the only source particles known which exist only 'in captivity' or 'confinement'. There are (as far as we know) no free quarks. They are all *permanently* bound together in baryons and in mesons.

Accepting the quark as a source particle within the framework of quantum field theory requires, however, the further assumption of appropriate glue particles that would provide the (elementary) strong force between them.

Table 12.3. *The source particles.*

Leptons	Baryons
Electron	Proton, neutron,
Muon	and many other
Tauon	heavier particles

Note: the muon lepton and the tauon lepton, are also called 'mu' meson and 'tau' meson (not to be confused with the mesons of Table 12.2 as these leptons have nothing to do with strong interactions).

These are called *strong gluons*. They perform the same task for quarks that photons perform for electrons: they mediate the force. Thus, the theory requires us to assume gluons if we assume quarks.

At this point there is an important difference between the quantum field theory of electromagnetic forces and the one of strong forces. The former, quantum electrodynamics (QED), has electrically charged source particles (positive or negative for particles, negative or positive for antiparticles). In the quantum field theory of strong forces the source particles (the quarks) have charges of *three* varieties and the antiparticles (the antiquarks) have the corresponding anticharges. In a moment of whimsy these three varieties of charge (*not electric* charge) were named 'color' such as red, green, and blue, and antired, antigreen, and antiblue. The quark field theory was therefore dubbed 'quantum chromodynamics' (QCD) in analogy to QED.

The gluons of QED, the photons, do not carry electric charge. But the (strong) gluons of QCD do carry color. Like electric charge, color determines whether the forces are attractive or repulsive: like colors repel, unlike colors attract. For some reason not fully understood, only color neutral objects can occur as free quantum particles. The red–antired combination of quark–anti-quarks is a meson, the three-quark combination of red, green, and blue quarks is the color neutral nucleon. A single quark can therefore not occur as a free particle. It is always confined, i.e. occurs in combinations with other quarks. Table 12.4 is the revised table of gluons.

If we now wish to list all the truly elementary particles (as we understand the subject at the current level of research) we must add to the table of glue particles a table of all the other particles including the source particles. Enumerating first the leptons or light particles we find that there exist in addition to the source particles already contained in Table 12.3 a number of neutrinos. These are electrically neutral particles of very small mass (possibly

Table 12.4. *The revised list of glue particles.*

Force	Name	Number	Spin
Electromagnetic	Photon	1	1
Weak	Weak gluon	3	1
Strong	Strong gluon	8	1
Gravitational	Graviton	1	2

Note: for all glue particles, the antiparticles are indistinguishable from the particles.

of no mass at all) which therefore move with nearly (or exactly) the speed of light. There are three of those, each associated with one of the lepton source particles. This association is observed in processes involving weak forces (leptons do not respond to the strong forces). For example, in a radioactive decay in which electrons emerge from a nucleus they would be accompanied by electron–neutrinos. In other processes where muons emerge they would be accompanied by muon–neutrinos. And one can show that those two types of neutrinos are really different particles. Thus, there are altogether exactly six leptons (Table 12.5).

The other kind are the quarks. They now take the place of the baryons of Table 12.3 as the elementary source particles. They are of course the source of the strong gluons (Table 12.4) and not of the mesons (Table 12.2). The quark hypothesis leads to six different quarks and all but one of them have by now been seen. By this is meant that one has seen hadrons which are made of these quarks since the quarks cannot exist as free particles.

The names of the six quarks are also given in Table 12.5. These are somewhat frivolous names indicative of the euphoria of the theorists who suggested them [12.11]. They do not celebrate the names or places of their discovery as has been the case, for example, with transuranic elements (see Note [12.7]). But in any case, this completes the two tables of particles at present considered elementary.

There are of course many other properties of these particles that we cannot pursue here. But one of these must be mentioned. It is the curious property that all the leptons and quarks have spin $\frac{1}{2}$ (in spin angular momentum units of $h/2\pi$), while the glue particles (Table 12.4) all have spins 1 or 2. This provides for considerable simplification and symmetry in the quantum field

Table 12.5. *Leptons and quarks.*

Six leptons		Six quarks	
electron	electron–neutrino	up-quark	down quark
muon	muon–neutrino	strange quark	charm quark
tauon	tauon–neutrino	top quark	bottom quark

Note: the antiparticles are all different from the particles.

theories describing these forces. In fact, one suspects that there is more symmetry behind the matter than we are presently aware of.

The number of quarks is seen to be the same as the number of leptons. Needless to say this has given rise to a great deal of speculation. It is the starting point of the Grand Unification Theory (Fig. 4.3). But the number of particles which we at present consider to be elementary has been pared down to a reasonable number. Let us not forget though that this is not the first time that such elementarity has been proposed. Bigger and better particle accelerators loom on the horizon [12.12], and experiments at much higher energies than have been available until now may well reveal entirely unexpected results. It is anyone's guess whether the quark theory will hold up and will become an established theory.

Annotated reading list for Chapter 12

Close, F. 1983. *The Cosmic Onion. Quarks and the Nature of the Universe*. New York: American Institute of Physics. A popular introduction to elementary particles representing our present-day understanding written by an active participant in this research program.

Morrison, P. *et al*. 1982. *Powers of Ten*. New York: W. H. Freeman. A delightful book of photographs with valuable comments by one of the most knowledgeable and successful popularizers among today's physicists.

Pagels, H. R. 1982. See the annotated reading list for Chapter 10.

Scientific American 1980. *Particles and Fields*. San Francisco: W. H. Freeman. A collection of well illustrated nontechnical reviews by leaders in that research field.

Weisskopf, V. F. 1979 *Knowledge and Wonder*. Cambridge: MIT Press. A great physicist presents a popular description of the natural world. Chapters 3 to 7 are especially relevant to the present topic.

Epilogue

We have reached the end of our journey. Of the many worlds of scientific quest we have visited only three, the world of high speeds, the world of gravitation, and the quantum world. These are the three worlds in which the greatest conceptual revolutions took place in our present century. Our journey has taught us how to look at the sophisticated science of today and how to evaluate the knowledge it provides so that we can construct our view of the physical world in the best possible way. Let us review what we have learned about scientific knowledge and about the world view it implies.

Scientific knowledge. It is healthy to be skeptical towards new results be they empirical or theoretical. Scientific theories undergo a long developmental stage with many trials and errors. Most theories before the nineteenth century were still in that developmental stage. Notable exceptions include Newtonian mechanics and gravitation theory. Even today there are established theories primarily in the physical sciences. Such theories involve models of nature which are but approximate and valid only within limits. These limits are known and provide a quantitative estimate of the error made. Within those limits, however, established theories involve laws that are akin to eternal truths. They range from such simple laws as that of Archimedes (that when one steps into a bath tub rim-full of water, the volume of the overflowing water is equal to the volume of the submerged body) to such sophisticated laws as the third law of thermodynamics (that there is a lowest temperature, absolute zero, at which matter has zero entropy [11.9]).

There are many different levels on which one can study nature. This can be seen best from the 'onion of matter' (Section 12b). But although one discovers a finer and finer structure, each level has its own importance, its own view of nature; there is no hierarchy of truth in these various levels. A certain holistic character pervades every level. At the same time, those levels must be consistent with one another in a limiting sense. They cannot be

incompatible or incommensurate; that would contradict the internal consistency and coherence of the structure of science; it would make every
established theory valid only until the next scientific revolution (Section 8b)
and make a mystery out of the proven cumulative nature of scientific
knowledge. Even the seemingly far apart worlds of classical physics and of
quantum physics are related in a limiting sense. There is after all only one
reality which is seen by us in different ways.

Each of these different ways is important in its own right each approximat-
ing nature in its own characteristic manner. Let us recall just two examples.
(1) As special relativity has taught us, time is not a public concept but a
private one: each reference frame has its own time, and there is no universal
simultaneity. Yet, on the level of Newtonian mechanics (i.e. within the
validity limits of that particular approximation) absolute simultaneity does
hold. And it does so to very high accuracy in many cases. (2) Einstein's
gravitation theory has taught us that space–time is curved, and that this
curvature is responsible for all gravitational motion. Yet, in Newtonian
gravitation (i.e. within a certain approximation) there is no such curvature
but there are instead universal forces responsible for the motion of the planets
as well as for the fall of an apple. These different concepts do not contradict
one another: one can prove that Newton's theory is an approximate form of
Einstein's.

Our world view. In accepting a scientific theory as true, one also accepts the
theoretical constructs in it. But do these constructs refer to things that are
actually there, or are they only devices of convenience invented to account for
what we observe in a logically consistent way? We have discussed two very
different views on this question, instrumentalism and realism.

Scientific realism makes two assumptions. First, there is a world 'out
there' that really exists and that is independent of our attempts to observe it
and in fact independent of our very being; secondly, scientific investigations
can make this world comprehensible to us. Instrumentalism questions the
first assumption and accepts as real only what is directly observable. All other
things such as electrons or quarks are theoretical constructs whose actual
existence is at least questionable.

In our everyday life we are all realists; we accept the world around us as
actually existing. But our attempts to carry this view into science soon
encounter difficulties. Our scientific investigations tell us that things are not
what they seem to be. We saw this in the above examples, first in the world
of high speeds and then in the world of gravity. Thus, even the classical world
is not quite what we believe it to be on the basis of our everyday experience.
But the greatest shock to our view of reality came from the quantum world.
And it is the development of quantum mechanics that led physicists to leave

classical realism behind and to embrace a more instrumentalist view (the Copenhagen interpretation).

But an extreme instrumentalism also encounters difficulties. Where, for instance, does the realism of 'direct observation' end and the questionable realism of 'indirect observation' begin? Is it the naked eye, magnifying glasses, an optical microscope, or an electron microscope? The historical development seems to have taken a dialectic road from naive realism to instrumentalism and is now settling on a critical scientific realism. This view includes quantum realism which accepts the existence of quantum particles identifying them by their mass, charge, spin and other properties that are always sharp. But quantum realism also accepts a new mode of existence, the *blurred mode*, for certain properties that are always sharp on the classical level but not on the quantum level: position, velocity, polarization, and others. A point electron can be in a blurred state of position, being *potentially* anywhere in a certain volume; interaction with a macroscopic object (which may be a measuring apparatus) actualizes one of these potentialities and places the electron at a particular location in that volume.

This view also accepts the probabilistic nature of the quantum world as intrinsic to its being (ontic indeterminism) rather than regarding it as a sign of our incomplete knowledge (epistemic indeterminism). It has been argued that this probabilistic nature is the result of our being part of nature and even that our consciousness plays a role in our description of the quantum world. Quantum mechanics tells us that the perturbation of the physical system by the apparatus used in our observations certainly plays a crucial role. But the observer as a human being does not affect the system; the creation of a permanent record does that. Scientific realism also believes that theoretical constructs (though not all) refer to actually existing things which are described differently on different levels of theory. Gravitation is really there whether it be described by forces or by space–time curvature.

We are fortunate to live in an age when science has reached such unprecedented heights. There is no better way to understand the world we live in than to take full advantage of everything our scientific quest is able to provide. If the present book has helped in providing the knowledge to do this, it has accomplished its task.

Notes by Chapters

Chapter 1

[1.1] In an ordinary microscope a light source illuminates the object one wants to see. In an electron microscope a beam of electrons does this job: the electrons are sent through the (very thin) object and then hit a screen much like a television screen. The object modifies the electron beam and therefore an image of its structure is visible on the screen.

[1.2] Sir Arthur Eddington, then the director of the Cambridge University observatory, became famous among philosophers when he posed the following question. 'Consider the table I am writing on. It is an object made of wood that a carpenter made and that everyone is familiar with. But it is also a complicated array of atoms and molecules that make up the cells of wood that in turn make up that table. Which is the "true" table?'

Chapter 2

[2.1] This is a very important concept. By 'model' one does not mean a scale model as one makes for airplanes or a replica. Reality has many faces and each of them is here believed to be important because each shows a different aspect of reality not fully conveyed by the others. In addition to Eddington's example (see Note [1.2]) the rainbow provides another instance. It has two 'faces' that are well known: the colored arc seen by everyone and the many water droplets in the air that spread the sunlight into its color spectrum; both faces are necessary for a complete description of the real object called 'rainbow'. These faces are the models of reality. Any one of them never tells the whole story but it tells a partial one which is true in its own right, different from the other models, and therefore indispensable.

[2.2] This is actually an oversimplification. While Fermat indeed stated his principle originally in this way (1657), he became well aware of the fact that it is not applicable in all cases. However, Fermat did not know how to resolve this difficulty in a satisfactory way. Not aware of this problem, some scientists and philosophers of his time (as well as of later periods well into the eighteenth century) used the minimum principle as an example of God's action in Nature. A revised principle which continues to carry Fermat's name and which is generally valid can, however, be stated. We shall not pursue this matter any further.

[2.3] In order to have specific examples of laws of nature these laws will be stated here in some detail. The law of reflection refers to a light ray incident on a mirror making an angle α with the line that is perpendicular to the mirror (the normal). The reflected light ray will make an angle β with respect to that normal. The law of reflection asserts that $\alpha = \beta$ (see Fig. 2.1). The law of refraction refers to the break of a straight light ray as it passes from one homogeneous medium to another, for example, from air into glass. If the two media are labeled 1 and 2 so that their indices of refraction are n_1 and n_2, and the angle with the normal (to the interface plane) of the ray in medium 1 is α_1 and of the ray in medium 2 is α_2, then the law of refraction (also known as Snell's law) asserts that $n_1 \sin \alpha_1 = n_2 \sin \alpha_2$, where 'sin' is a trigonometric function (see Fig. 2.2).

Chapter 3

[3.1]   This brief statement on the scientific method refers only to its *prescriptive* aspects: how science should be done. The scientific method in its *descriptive* aspects, how scientists go about their business, has been studied primarily by philosophers and by historians of science. As such it is being revised in every new era of science. It raises questions about such activities as inductive reasoning, forming of hypotheses, deductive reasoning from such hypotheses, and confirmation of theories. Discussion of these matters would lead us too far afield. But we shall of course encounter specific instances of all of these.

[3.2]   It was only many years later when the phenomenon of destructive interference of light was discovered (Thomas Young, 1802) and when it became possible to explain thereby the colors of thin films, that the corpuscular theory had to be abandoned in favor of the wave theory of light. There was no way in which the corpuscular view could account for destructive interference.

[3.3]   In the 1930s when Lysenkoism began in the Soviet Union and then rose to become the accepted theory of the party, scientists who held different views, some of them distinguished geneticists who were respected internationally, were relieved of their posts. One of them, Vavilov, died in a Siberian labor camp. It was a time where scientific truth was not decided by free exchange of views among scientists but was dictated by the politics of the country. After Stalin's death Lysenko fell into disfavor and Soviet biology returned very slowly to the acceptance of those theories that had the support of the geneticists in the rest of the world. (See Martin Gardner, 1957.)

[3.4]   This expression indicates 'less than certainty but not far from it'. A conjecture or hypothesis is considered likely to be correct if it has been confirmed repeatedly. Confidence increases with the amount of confirmatory evidence. But no amount of repetition of confirmation can turn that hypothesis into a certainty; and if certainty cannot be assured the matter must be left at the level of very high probability. We cannot be certain that the sun will rise tomorrow at the expected time because we cannot rule out with certainty an unforeseen cosmic catastrophe that may interfere with the normal course of events. But it is overwhelmingly probable that it will rise on schedule.

[3.5]   Geometrical optics was discussed briefly in Section 2a. What electromagnetic theory is all about will become clearer in Section 6a.

[3.6]   Phlogiston was believed to be a 'substance' that entered into metal oxides (then known as calxes) upon heating and that thus produced metals. Conversely, phlogiston was believed to leave a metal when that metal was changed back to calx. When Lavoisier in the second half of the eighteenth century carried out quantitative experiments it became clear that phlogiston would have to have a negative mass. Since mass must be positive the phlogiston theory had to be abandoned. It had been based on insufficient empirical evidence. Lavoisier discovered oxygen and showed that calxes are metal oxides (chemical compounds of a metal with oxygen) and that they are heavier than the metals from which they are produced.

[3.7]   This great discovery by two young physicists in the United States who are both natives of China, Tsung Dao Lee and Chen Ning Yang, earned them the Nobel prize in physics within the same year (see I. Asimov, 1972).

[3.8]   The mathematical notion of a group is not very difficult to understand. Since we shall encounter it several times we shall define it here in general terms. Consider a set, a collection of items called 'elements'. There may be a finite number of them (finite group), or an infinite number (infinite group). To have something specific in mind take the set of all integers (both positive and negative) as an example. Assume that there also exists some 'operation' that can be performed with any ordered pair of these elements. In our example this would be the operation of addition. (The ordering of the pair is not necessary in this particular case of addition since that operation is commutative. Not all operations are.) This set together with the operation is called a group provided the following three conditions are satisfied:

(a) The operation turns any ordered pair of elements of the set into something that is also an element of the set. In our example the addition of two integers is also an integer.

(b) The operation is associative. If one denotes the operation by 'o' then this associative property can be expressed by $(aob)\ oc\ =\ ao\ (boc)$.

(c) The set contains an element (called the identity and usually denoted by 'e') so that the operation on it and some other element (no matter in which order) reproduces that other element. Symbolically, $eoa\ =\ aoe\ =\ a$. In our example the identity element is the number 0 because adding it to any integer reproduces that integer.

(d) For every element in the set there exists exactly one other element (called the inverse) with the property that the operation on that pair of elements (no matter in which order) gives the identity element. In our example the 'inverse' of an integer is the corresponding negative integer (the inverse associated with 17 is -17) because addition of the two produces 0, the identity.

If the ordering of the pair does not matter for the outcome of the operation (as in the case of addition) the group is called 'abelian' after the mathematician Niels Henrik Abel. In the case of the symmetry of a square the elements of the group are all the multiples of rotation by 90 degrees; the operation is the combination of two rotations. It is easy to verify that these form an abelian group.

Chapter 4

[4.1]   Empiricism is the philosophical view which claims that all human knowledge (with the possible exception of logic and mathematics) is based on experience. Specifically, our sense experiences provide justification for our belief in the correctness of our knowledge. This view is therefore opposite to rationalism. The best known empiricists were John Locke (1632–1704), David Hume (1711–76), and Bertrand Russell (1872–1970).

[4.2]   Stimulated emission is the process by which an atom in an excited state (i.e. not in its ground state) is forced to emit radiation due to a surrounding of radiation of appropriate frequency. This process is used in the generation of laser radiation. Since this phenomenon was already known in 1917 it is not to the credit of physicists and engineers that the first laser was only constructed as late as 1960. The appropriate technology could have been developed sooner.

[4.3]   The notation 10^n is in common usage in mathematics. It is an abbreviation for the number that starts with 1 and is followed by n zeros. Thus, $10^3 = 1000$ and $10^0 = 1$. The notation 10^{-n} means that one should take the reciprocal of the number 10^n. Thus, $10^{-3} = \frac{1}{1000}$ and $10^{-1} = \frac{1}{10}$.

[4.4]   It is remarkable that scientists were so confident of the correctness of the theory of the unification of electromagnetic and weak forces that they awarded the Nobel prize to the three main architects of that theory, Sheldon Glashow, Abdus Salam, and Steven Weinberg, in 1979. That was five years before the gluons W and Z predicted by that theory were first seen by experimenters. The discovery of those particles yielded the Nobel prize for Carlo Rubbia and Simon Van der Meer in 1984. The electroweak theory is now considered established.

Chapter 5

[5.1]   In 1632, at age sixty-eight, Galileo published his book *Dialogue on the two Chief Systems of the World* in which he maintained the superiority of the Copernican over the Ptolemaic system. But Copernicus' view puts the sun in the center of the world and has the earth move about it rather than having the earth in the center as Ptolemy thought. The belief that the earth is moving, however, was 'false and altogether opposed to the holy Scripture', according to the authorities of the Church at that time. Galileo was therefore summoned to Rome by the Inquisition. There, he was forced to recant, and he was sentenced to incarceration. His sentence was mercifully commuted to house arrest in Florence where he remained until his death in 1642.

[5.2]   Uniform motion is motion with constant velocity which means constant speed and fixed direction of motion.

[5.3]   It is interesting that many of the results reported in this work were known to Newton already some 20 years earlier when he was still in his mid-twenties. These were in fact his most productive years. But he had some difficulties with the gravitational force law for spherical objects which he was not able to resolve until just a few years before the publication of the *Principia*. A second edition of this work appeared in 1713. It contained among other additions a more explicit statement of his belief in a Deity and its relation to the grand design of the world whose laws of motion he recorded mathematically in this work. A third edition appeared in 1726 just one year before his death. The first translation into English was not published until 1729. It is best known in its revised version by F. Cajori (1934).

[5.4]   The *center of gravity* of a body can be understood as follows. If a body moves under the influence of a force it will follow some complicated orbit. Now there is a remarkable theorem: exactly the same complicated orbit would be traversed if that force were acting on just one single point which is exactly as heavy (has the same mass) as that body provided that point is suitably chosen. Such a point is called the center of gravity. For symmetrical objects the center of gravity is the point of symmetry. For example, the center of gravity of a ring is at the center of the ring. This shows that the center of gravity is not necessarily inside the matter of the body; it can also be 'in the air' as in the case of the ring. Of course, since the center of gravity is a point, it traces out a line as it moves along, while a body which has extension will trace out an orbit with a cross-section of some size. In many problems of physics, for example in the description of the solar system, the distances between the bodies (planets and sun) are so large compared to the sizes of the bodies that these bodies can be approximated by points. But these points must be given a mass so that the force law of gravitation can be applied. When one replaces the planets and the sun by 'massive points' in this way one must place these points at the respective centers of gravity of these bodies.

[5.5]   The concept of a group was defined in Note [3.8]. Its application to uniform motion will be seen shortly: it is the group of 'boosts' discussed below.

[5.6]   This link between invariance properties of the laws of motion (under transformations from one inertial frame to another) and the conservation laws of mechanics is indicated by several examples in Table 5.1. The translation transformation of time which is listed there refers to the shift of the point of time from which time is being measured. For example, one can measure time from the year 0 AD (Christian calendar) or one can measure it instead from the year 622 AD (Muhammadan calendar). Clearly, such a change cannot make any difference to the laws of motion; they must therefore be unchanged (invariant) also under such transformations.

Chapter 6

[6.1] The very word 'electric' derives from the Greek word 'electron' which means 'amber'. When rubbed, amber was known to attract sufficiently light objects. Today we know that such rubbing creates static electricity.

[6.2] By Maxwell's time Newton's corpuscular theory had been dead and Huygens' wave theory of light had triumphed as a result of new discoveries in the early nineteenth century especially by Thomas Young (see Note 3.2).

[6.3] After the accuracy of the measurement of the speed of light had reached about one part in 100 million it was decided to abandon the old standard for the meter and adopt the speed of light in a vacuum as a new standard. It was then agreed upon to *fix* that speed to be the above stated number and to define the meter in terms of it. The new definition of the meter is now the length which light travels in a vacuum during $\frac{1}{299792458}$ th of a second.

[6.4] When a star is just overhead so that the light would arrive vertically if the earth were at rest relative to the sun, the small angle by which the telescope must be tilted because of the motion of the earth can be computed from the ratio v/c where v is the speed of the earth and c the speed of light. If one denotes the very small tilt angle by α then $\alpha = v/c$. On the other hand, when the star is not just overhead, the formula is a little more complicated.

[6.5] Michelson developed an extremely sensitive apparatus called an 'interferometer'. It could measure differences in the time of arrival of two light rays to an accuracy of one part in ten billion. He accomplished this by having the two light rays start from the same source, travel different paths, and finally come together again interfering with one another. In the experiment with Morley one ray traveled *perpendicular* to the motion of the earth to a mirror and back, the other one traveled *along* the direction of motion of the earth to a mirror and back.

[6.6] The year 1905 is called Einstein's miracle year, annus mirabilis, because during that year there appeared in print five scientific papers authored by Einstein. Four of these were of such high quality and far-reaching consequences that each of them would have been deserving of a Nobel prize. For one of these which contains the theory of the photoelectric effect he actually did get that prize (in 1922). Another one contained the foundations of the theory of special relativity; in a third one he derived the famous relation $E = mc^2$.

[6.7] This is strictly true only for propagation in a vacuum. Propagation in air involves a speed only very slightly different. Quite generally, however, propagation of light in any medium, air, water, or glass, is independent of the motion of the source. It does depend on the motion of the medium (see Note [6.12] below).

[6.8] From Fig. 6.4 follows

$$\frac{t}{t'} = \frac{D}{d} = \frac{D}{\sqrt{[D^2 - (\frac{1}{2}vt)^2]}} = \frac{D}{\sqrt{[D^2 - (vD/c)^2]}} = \frac{1}{\sqrt{[1 - (v/c)^2]}}.$$

[6.9] In those accelerators another particle called a 'pion' is produced first. That particle then decays into other particles one of which is the muon.

[6.10] This decay process is governed by the laws of quantum mechanics (Chapters 10 and 11). 'Lifetime' refers to an average as does the life expectancy of people. Particles, like people, do not die all at the same age. There exist different probabilities for dying at different ages. Quantum mechanics provides these probabilities for particles. But the important difference between the lifetimes of people and of particles is this: for people the life expectancy clusters around some most probable age; for particles the range of ages is much larger and cannot be specified in quite the same way. One knows the *half-life* which is the time it takes for half of the sample to decay. After two half-lives only one-quarter of the original sample survives, after three half-lives only one-eighth, etc. For muons the half-life is 2 microseconds (2 millionths of one second).

[6.11] If one observes a luminous cube that moves past in a direction parallel to one of its sides, one may expect to see it contracted in the direction of motion so that the side facing the observer appears as a rectangle rather than as a square. However, as it moves by, the observer will also receive some light from the sides perpendicular to the direction of motion. The amount of that light can be shown to compensate exactly for the contraction. It will therefore look like a square despite the contraction (see Weisskopf 1960).

[6.12] The speed of light in water at rest is c/n where n is the index of refraction of water. But when the water is moving with speed v in the same direction as the light ray then the formula for the relativistic combination of speeds (obtained above) gives

$$u' = \left(\frac{c}{n} - v\right)\Big/\left(1 - \frac{v}{nc}\right).$$

The right-hand side is approximately equal to $u - v(1 - 1/n^2)$ (in the approximation where $v^2 \ll c^2$). And that is just what was observed by Fizeau within the accuracy of this approximation.

[6.13] Since speed is measured in m/s (meter per second) and acceleration is the change of speed per second, its units are m/s^2.

[6.14] A derivation of this formula in full generality is beyond our scope. However, we can do it in a special case where we can use Newtonian mechanics together with some knowledge about photons. These 'quanta of light' will be discussed later (p. 131); but all we need to know about them here is that they behave like particles with energy hf and momentum hf/c, where f is the frequency of light and h is a constant.

Consider the following 'thought experiment'. Two photons are emitted simultaneously from a source which remains at rest during the emission. We shall describe that source by means of Newtonian mechanics. If the source remains at rest in reference frame R while the two photons are being emitted, they must each have the same energy hf and the same momentum hf/c, and they must move in opposite directions. This is required by the law of conservation of momentum. Conservation of energy requires that the source loses energy in the amount of $E = 2hf$.

But let us now look at this process from another reference frame, R', in which the source moves with speed v in the same direction as one of the photons. A frequency f (as measured in frame R) is seen as a higher frequency when the source approaches and as a lower frequency when the source recedes. This is known as the Doppler effect. Quantitatively, the new frequencies are $f' = f(1 + v/c)$ and $f' = f(1 - v/c)$, respectively. Therefore, relative to R' energy conservation requires that the source loses energy E' of the amount $\qquad E' = hf(1 + v/c) + hf(1 - v/c) = 2hf.$

But this is exactly the same amount as lost when seen in reference frame R. That energy loss therefore does not depend on the speed v. The law of conservation of momentum requires that the loss of momentum by the source, p, is

$$p = (hf/c)(1 + v/c) - (hf/c)(1 - v/c) = 2hfv/c^2.$$

But the momentum loss of the source is $p = mv$ according to Newtonian mechanics. Therefore, the source must have lost an amount of mass m,

$$m = 2hf/c^2 = E/c^2.$$

This is the equation we wanted to obtain. It relates the energy loss to the mass loss by $E = mc^2$ and the energy loss is indeed independent of v. The lost energy is mass energy.

This thought experiment can be carried out for any frequency f, except that this frequency cannot be so large that the corresponding energy of each photon, hf, requires the source to lose more mass energy than the source has to begin with. The largest frequency is such that *all* the mass energy of the source is lost. In that case, E is the *total* energy of the source and m is its total mass which the source had before emission of the photons. Thus, $E = mc^2$ also holds for that situation.

If the same argument is carried out using the formula for the Doppler effect according to special relativity theory (rather than according to Newtonian physics as was done above), the total energy of the source as it is seen to move in R' with speed v is found to be related to the mass m which it has when at rest in a very similar way:

$$E = \gamma mc^2 = \frac{mc^2}{\sqrt{[1 - (v/c)^2]}}.$$

The symbol γ was defined on p. 66. This is the *total* energy of any object of mass m. It consists of the energy when at rest $(v = 0)$, mc^2, and the kinetic energy according to special relativity

$$K = mc^2(\gamma - 1).$$

When K is written out for $(v/c)^2$ small compared to 1,

$$K = mv^2/2$$

which is exactly the kinetic energy according to Newtonian mechanics. This is another example of the smooth mathematical transition of the formulas of special relativity to the formulas of Newtonian mechanics in the approximation in which $(v/c)^2$ is small compared to 1 (compare pp. 61–2).

[6.15] The nuclei of atoms consist of only two kinds of elementary particles, *protons and neutrons*. The chemical properties of an atom are determined by the number of protons in its nucleus. The uranium atom has a nucleus containing 92 protons. Chemical elements usually occur in nature as mixtures of atoms whose nuclei differ in the number of neutrons they contain. These different kinds of atoms belonging to the same chemical element are called *isotopes*. Naturally occurring uranium has three isotopes, 234, 235, and 238, containing 142, 143, and 146 neutrons, respectively (and 92 protons each). Most of it is 238, only 0.7 per cent of it is the isotope 235, and the amount of 234 is even much smaller than that.

[6.16] In a Euclidean space the hypotenuse of a right angled triangle obeys the law of Pythagoras: if x and y are the two perpendicular sides, the square of the third side (the hypotenuse) is $x^2 + y^2$. Now in Minkowski space x and ct are also in a sense 'perpendicular'. But the 'triangle' they form has a third side which is $x^2 - (ct)^2$. A minus sign instead of a plus sign makes for the difference between these two spaces.

[6.17] This limitation is not an assumption added to the theory but emerges from the two postulates, Section 6c. These imply the formula for the way in which speeds add, Section 6d. As we saw, that formula does not permit one to increase the relative speed of light by moving in the opposite direction. Special relativity also leads to the conclusion that the energy of a body moving close to the speed of light is extremely large: in Note [6.14] above that energy was found to be $E' = \gamma mc^2$ and the factor γ becomes extremely large when v/c is almost 1. An additional increase in speed would require a huge amount of energy; if one wants to increase the speed to exactly that of light an infinite amount of energy would be needed. This is clearly physically impossible.

On the other hand, it is mathematically not excluded that objects move with a speed greater than light. But such objects can never slow down enough to reach the speed of light; they would always have to move faster than light. Such faster-than-light particles are called 'tachyons'. Some theoretical physicists have speculated whether fundamental particles of that nature might exist. Experimentalists have looked for them but have found none despite considerable efforts. In the absence of such particles there is no way in which a signal (which requires energy) can be sent faster than the speed of light.

Chapter 7

[7.1]   The three laws discovered by Kepler are descriptive of the motion of the planets: (1) they revolve in ellipses with the sun in one of the focal points; (2) they move faster when closer to the sun so that the line connecting a planet to the sun sweeps every day over the same amount of area; (3) the ratio of the periods (time to complete one revolution) of two planets T_1/T_2 and the ratio of their average distance from the sun D_1/D_2 are related by the formula $(T_1/T_2)^2 = (D_1/D_2)^3$. Newton used these laws to find a force of gravity so that these laws would emerge as a consequence.

[7.2]   The story of an apple having fallen on Newton's head and having caused him to discover the law of gravity is unfortunately pure fiction. A similar fiction is the story of Galileo having dropped objects from the leaning tower of Pisa.

[7.3]   As an example, consider the weight of an apple on the surface of the earth. It is just the force of gravity that acts between the apple and the earth. Substituting into Newton's formula, we put m_1 equal to the mass M of the earth, m_2 equal to the mass m of the apple, and d equal to the distance between the center of the earth and the center of the apple which is just the radius R of the earth since the radius of the apple is negligible compared to R. The result is

$$G\frac{Mm}{R^2}.$$

If we abbreviate the expression GM/R^2 by the symbol g then the weight is simply gm. This identifies g as the *acceleration of gravity* on the earth's surface (9.8 m/s^2).

[7.4]   The lecture was delivered in German and there was no manuscript. It was translated simultaneously to the Japanese audience by a Japanese who had studied in Germany and notes were taken in Japanese. This paper (Einstein 1982) is a translation of these notes. While the record of this lecture by Einstein is thus only third hand, Einstein's own writings of a later date as well as interviews with him confirm the story told there.

[7.5]   It is understood here that the different coordinate systems that are equally good are all Gaussian, i.e. that each of them provides an unambiguous specification of every point in space–time.

[7.6]   The perihelion is a planet's point of closest approach to the sun.

Chapter 8

[8.1]   The holistic feature of the biological sciences was strongly advocated by Michael Polanyi (see Reading List).

[8.2]   The problem of theory change in the philosophy of science received an important contribution from the historian of science Thomas Kuhn. His book *The Structure of Scientific Revolutions* (see Reading List) was widely studied and influenced not only the physical sciences but also the biological and social sciences. His followers try to answer the problem of theory change by studying how the scientific community comes to 'abandon one time-honored way of regarding the world and of pursuing science in favor of some other, usually incompatible, approach to its discipline'. Others go even farther in that direction and expect to learn to understand scientific revolutions by studying the sociology of the scientific community and the forces that influence it. Another approach is based on scientific realism concentrating more on the subject matter of science than on the scientists. Our view is in closer harmony with the latter.

[8.3]   Studying the behavior of the group rather than of the individual leads to holistic conclusions in psychology, sociology, and other social sciences.

Chapter 9

[9.1]   Statistical mechanics is the theory which accounts for the mechanics of systems composed of a very large number of particles by averaging the mechanical properties of the individual particles and their interactions. For example, the relations between the pressure of a gas, its volume, and its temperature, are obtained from the masses and velocities of the individual molecules of the gas by suitable statistical considerations.

[9.2]   Rutherford used a beam of electrically charged particles called 'alpha particles' that are emitted from certain radioactive elements and that are much heavier than electrons. When he directed that particle beam onto thin foils of various metals, he observed that almost all particles seem to encounter no resistance at all and passed through the foils without deflection. A few, however, were deflected strongly, even in the nearly backward direction. This indicated that each atom presented a very heavy obstacle which is, however, small compared to atomic size. Most of the atom thus seems to be empty or nearly empty and all of the atomic matter appears to be concentrated in a very small region of space. This small region became known as the nucleus of the atom. The atomic electrons surround the nucleus but are so light that they offer no resistance to the heavy incident alpha particles.

[9.3]   This term may be confusing since part of that radiation is in the visible region of the spectrum and can be very bright. An ideal black surface absorbs all radiation that it receives, reflecting none. But the better a surface absorbs radiation the better it is able to emit it. And an ideal absorber is also an ideal emitter. Therefore, a body that is ideally black (for incident radiation) makes at sufficiently high temperature for an ideal emitter and its radiation is then black-body radiation. Examples of approximately 'black' bodies include the sun (!) and a potter's kiln as mentioned before.

[9.4]   A Joule is the metric unit of energy named after the nineteenth-century English physicist James Prescott Joule. If energy is produced at the rate of 1 Joule per second one has a power generation of 1 Watt which is a more familiar quantity. The units of the constant h are energy times time, a quantity usually called 'action'. For this reason h is also known as 'the quantum of action'.

[9.5]   Despite the revolutionary importance of Bohr's atomic model it is not based on sophisticated mathematics but can easily be understood with a little algebra. A hydrogen atom consists of a proton and an electron. The proton is its nucleus. It has the same amount of electric charge as the electron but is positive. The force between the proton and the electron is therefore attractive. It is given by Coulomb's law,

$$F = e^2/r^2,$$

where r is the distance between the two particles. The units are so chosen that the constant in front of this law is just 1 (Gaussian units) which is most convenient here. Since the proton is about 2000 times heavier than the electron one can assume that only the electron moves and that the proton remains at rest at the center (in good approximation). If the electron's orbit is a circle around the proton then the distance r does not change nor does the magnitude of the electron's velocity, v. The acceleration which the proton exerts on the electron is directed toward the center and is according to Newton v^2/r. His force law (force equals mass times acceleration) then reads

$$e^2/r^2 = mv^2/r.$$

The orbital angular momentum L of a body revolving about a center is by definition the product of its linear momentum (magnitude mv) times the distance of the line along which the momentum acts from the center. In our case it is just mvr. It is constant provided the moving electron does not radiate. Bohr assured the lack of radiation by *postulating* L to be a constant. He furthermore assumed it to be proportional to h, namely $nh/(2\pi)$. Thus, by assumption

$$mvr = nh/(2\pi)$$

where n is a positive integer. This is the crucial *postulate* that Bohr made and that assured the stability of his model of the atom. Together with the above force law one now has two equations for the two unknowns v and r. They can easily be solved by algebraic manipulations and one finds

$$v = c\alpha/n,$$
$$r = n^2h/(2\pi\alpha mc).$$

The Greek letter α is a shorthand for the often occurring dimensionless combination $2\pi e^2/(hc)$ which is approximately $1/137$ when the numerical values of the constants are substituted. One has now obtained both the speed of the electron and its orbital radius. The latter gives the size of the atom.

The total energy of the atom consists of the sum of the kinetic and the potential energy,

$$E = mv^2/2 - e^2/r,$$

where the second term is the potential energy of the electron in the electric force field of the proton. Substitution of the formulas for v and r obtained above leads to

$$E_n = -\tfrac{1}{2}\alpha^2mc^2/n^2.$$

The subscript n on E indicates the particular size of the angular momentum. The negative sign can be understood as follows. When the proton and the electron are infinitely far apart ($r = \infty$) and at rest ($v = 0$) the formula for the total energy E above gives $E = 0$. The energies of the atom are measured relative to that state having no energy. In order to separate the hydrogen atom into this state of infinitely distant particles work must be done against the Coulomb attraction. Thus one must *add* to the energy of the atom to bring the system to a state of zero energy. Hence, the energy of the atom must be negative.

The state of the atom which has the lowest possible energy (the 'ground state') is the one with $n = 1$ because that is the value of n for which E is most negative. For this state one finds from the above formulas $v/c = \tfrac{1}{137}$ and $r = \tfrac{1}{2} \times 10^{-10}$ meter.

If the electron jumps from a state characterized by n (energy E_n) to one of larger energy (E_m) characterized by m, where $m > n$, then a photon of energy hf must be absorbed in order to ensure the conservation of energy,

$$hf_{mn} = E_m - E_n.$$

This is the famous *Bohr frequency condition*. It relates the frequency of the emitted or absorbed radiation to the energies of the initial and final states of the electron in the atom. If the final energy is larger radiation has been absorbed. The frequency of the absorbed radiation follows from the last two equations to be

$$f_{mn} = R(1/n^2 - 1/m^2).$$

R is a constant that depends on the fundamental constants e, h, c, and m. It is called the Rydberg constant. This formula for the Rydberg constant agrees extremely well with the observed value. The ratio of calculated to observed values is 1.000 54. Similarly, all the frequencies of radiation emitted or absorbed by hydrogen were found to be in excellent agreement with the above formulas. The Bohr model was a big success.

Chapter 10

[10.1] It is of course unfair to associate a profound intellectual development, especially one of this magnitude, with only a very small number of prominent people. Their interaction with colleagues whether directly or via written communication usually contributes to the eventual discoveries. Such contributions may be significant but are often nearly impossible to evaluate. In addition to the five founders of quantum mechanics already mentioned, others who contributed in important ways can be added. These include in alphabetical order Albert Einstein, Pascual Jordan, and Wolfgang Pauli. A fair and complete list of credits would be very difficult to construct.

[10.2] The alert reader may wonder whether the discovery of the photon was not a step backward to the old Newtonian corpuscular theory of light that had been completely discredited by experiments early in the last century (see Note [3.2]). This is, however, not at all the case. Newton's corpuscles were classical particles while the photons are very much quantum-mechanical particles. As we shall see, the photons are not in contradiction with Maxwell's classical electromagnetic theory although this may seem to be so at this stage. Newton's model of light, however, *is* in contradiction with that theory.

[10.3] For this and related episodes in the history of quantum mechanics see Jammer 1966 or Pais 1986.

[10.4] The phenomenon of diffraction that was first studied with visible light must hold for all electromagnetic radiation according to Maxwell's theory. The only thing that matters in producing this phenomenon is the ratio of the width of the slits in the baffle to the wave length. That ratio cannot be much larger than 1. Thus, if one were to use X-rays which have a much shorter wave length than visible light, one must use a baffle with much narrower slits than can be manufactured. Max von Laue had the brilliant and seminal idea (1912) to use a crystal as a baffle: the spaces between the atoms are of just the right size, i.e. they are of approximately the size of X-ray wave lengths. His successful experimentation verified this conjecture and provided the foundations for a new field of study, X-ray crystallography. The resultant diffraction patterns provide a great deal of information about the structure of crystals.

[10.5] In discussing interference it is convenient to use the notion of *phase*. Any two points along the axis of Fig. 10.5(a) that are separated by exactly one, or two, or three, etc. wave lengths are said to be 'in phase'. At two points that are in phase, the wave behaves the same way; they have equal amplitudes. When two waves of the same wave length come together in phase they interfere *constructively*. But if two points along the axis are separated by $\frac{1}{2}$ wave length (or $\frac{3}{2}$, or $\frac{5}{2}$, etc.) they are called 'out of phase'. At those points the wave has amplitudes that are of the same magnitude but of opposite sign. When two waves of the same wave length come together out of phase they interfere *destructively*.

[10.6] There is a mathematical complication here in that this quantum mechanical amplitude ψ must be characterized by *two* numbers rather than only one as is the case for ordinary waves. If these two numbers are called 'a' and 'b' then by the square is meant here $a^2 + b^2$. Such pairs of numbers are known in mathematics as 'complex numbers'. To indicate that difference one writes for the square of ψ not ψ^2 but $|\psi|^2$. This technical point will not concern us very much.

[10.7] One characteristic of the quantum world is the fact that apparently reasonable questions cannot be answered empirically *in principle*. Consequently, we shall at times run into questions that would be meaningful in the classical world but that are meaningless in the quantum world. An example of such a question is the following: when an electron in an atom jumps from a higher energy orbit to a lower energy one, the atom emits a photon. What is the mechanism that creates this photon?

[10.8] The hidden variables interpretation is one of several interpretations which are unpopular with the vast majority of working scientists in the field. However, as long as there does not exist conclusive invalidating *proof*, it is a matter of scientific fairness and objectivity that this interpretation not be discarded.

[10.9] What has been said about the hidden variables interpretation (see the preceding Note [10.8]) also applies to the statistical interpretation.

[10.10] For an excellent discussion of this so unexpected quantum behavior see R. P. Feynman, R. B. Leighton, and M. Sands 1965.

[10.11] The uncertainty principle has been stated in Section 10d in the context of massive quantum particles. For these the width in momentum Δp of the probability distribution is equivalent to a width in velocity Δv because $p = mv$. However, that principle also applies to photons which are massless quantum particles. In that case it is correct only when stated in terms of Δp. Photons move exactly with velocity c (in a vacuum) and there is no probability *distribution* in velocity. For photons the relation $p = mv$ is meaningless since $m = 0$; it is replaced by $p = E/c$ where E is the photon energy hf.

Chapter 11

[11.1] That the electron is a point particle is an experimental fact. We know the force law which holds outside the electron. That is the law which tells how the electromagnetic force exerted by the electron on other charged particles changes with distance. It can be tested with great accuracy, and it was found to be valid down to distances of at least 10^{-17} m. Even down to these extremely short distances one is still 'outside' the electron. No indication of 'crossing into the interior' is noticeable.

[11.2] In fact, the number of observables needed for a complete specification of a quantum system is just half that for a classical one. This can be understood as follows. In classical mechanics one needs all the positions and velocities for a complete solution of the deterministic equations of motion. But the quantum mechanical observables of positions and velocities are such that, while all positions are compatible with one another and all the velocities are compatible with one another, a position is not compatible with a velocity if both refer to the same particle in the same direction (this is of course closely related to the uncertainty relation). Therefore, one complete set of compatible observables in quantum mechanics is the set of all position observables; another would be the set of all velocity observables. In either case their number is only half the number of all the observables needed in the classical case.

[11.3] In such a complete set each observable must be compatible not only with the energy observable but also with all the others. Two observables that are both compatible with the energy observable but not with each other belong to different complete sets.

[11.4] The general superposition principle states that if the quantum system can either be in state ψ_1 or in state ψ_2 and if the probability for being in the former is 70 per cent, say, and for being in the latter 30 per cent, then the ψ for the system is $a\psi_1 + b\psi_2$ where $a^2 = 0.7$ and $b^2 = 0.3$. In our case $a = b$ since the probabilities are equal.

[11.5] The loss of information as a result of the large numbers of processes taking place that cannot be followed in detail is responsible for this irreversibility. Such irreversibility is well-known in classical macroscopic processes, for example those associated with heat flow: if one part of a body is hot and the other one is cold the temperature tends to distribute uniformly over the whole body. The opposite process does not take place. The process is irreversible (unless external influence is applied). Such irreversibility is measured by a quantity called 'entropy'.

[11.6] This is actually an experiment on a pair of photons rather than electrons. But their polarization behaves very similarly to the way electron spins behave. It was performed in Paris during 1981–82 by Alain Aspect and his collaborators J. Delibard, P. Grangier, and G. Roger. See Rohrlich 1983.

[11.7] There is a special law in quantum mechanics that ensures this matter. It is called 'superselection rules'.

[11.8] The original work by Birkhoff and von Neumann was developed further by quite a number of mathematicians and physicists. Various generalizations were necessary in order to make quantum logic exactly equivalent to the mathematics of quantum mechanics as we know it today.

[11.9] Entropy (see Note [11.5]) is used as a measure of disorder. It is so chosen that at absolute zero of temperature its value is zero. The famous second law of thermodynamics states that the entropy of an isolated system can never decrease. As matter is being cooled and acquires a lower entropy, it is clearly not isolated and the entropy of its surroundings (the apparatus that cools it) must necessarily increase.

[11.10] When the heat capacity of helium is measured as temperature increases through the lambda point, one finds a strange curve: it first increases, then decreases very rapidly, and then increases again. That curve resembles the Greek letter 'lambda'. Hence the name.

Chapter 12

[12.1] The electric and the magnetic field are examples of fields (Section 6a). Another example is the velocity field of a fluid: a moving fluid (a liquid or a gas) will in general have different velocities at different points; the velocity field is simply the whole space occupied by the fluid with a little arrow at each point of the fluid indicating the direction of the velocity at that point (arrow direction) and its size (arrow length).

[12.2] As an example of this accuracy consider the electron. Because of its spin this electrically charged particle is also a little magnet. The strength of this magnet can be measured to an accuracy of better than 1 part in 10^{11}. Calculations by means of quantum electrodynamics have been carried out with great effort to the same accuracy. And theory and experiments agree in all 11 figures within the measured error.

[12.3] Cosmic rays (or cosmic radiation) bombard the earth at all times. They come primarily from the sun but also from outside the solar system. They are very high energy particles including photons, electrons, and mostly protons. As they collide with the atoms and molecules of the earth's atmosphere, further (secondary) particles are produced by their impact.

[12.4] While the grand unification is far from being successfully completed there seems to be little doubt that such a unified theory predicts the instability of all matter, even matter (unlike radioactive substances) that has been considered as stable until now. Of course such a decay would proceed extremely slowly. In quantum terms it means that the probability for such a decay is extremely small but not zero. Typically, a proton would have a 50–50 chance to decay if one could wait 10^{32} years which is about 10^{12} times as long as the age of the universe. This would not make it unobservable because it means that out of 10^{32} protons 1 is expected to decay every year (in average). Or, out of 10^{34} protons 1 will decay every three days. Large quantities of water contain very large numbers of protons (nuclei of hydrogen atoms). Experiments using suitable quantities like that are in fact in progress at the present time. No proton decays have been observed thus far.

[12.5]   In the present section the powers-of-ten notation will be used extensively. This notation is explained in Note [4.3].

[12.6]   A chemical compound is a pure substance, i.e. one that is not a mixture of several substances and that could therefore not be broken down into other substances by *physical* means such as gravitation, magnetism, etc. An organic chemical compound is one that contains the element carbon.

[12.7]   At the time of Dmitri Ivanovich Mendeleyev (1834–1907), the discoverer of the periodic table of elements, only 92 elements were thought to exist, the last one, 92, being uranium. Since that time heavier elements which fit into that table beyond uranium have been created artificially. They are called 'transuranic' elements. Their lifetime is very short since they decay rapidly into lighter elements. They have been named at first after the planets beyond Uranus (93 is neptunium, 94 is plutonium) and then after people and places (95 is curium after Curie, 99 is einsteinium, 95 is americium, 98 is californium, etc.). The periodic table has thus been extended to over 100 elements.

[12.8]   The nuclei that belong to different elements differ in their electric charges: element n has a nucleus with exactly n times the charge of the proton. But for most elements the nuclei come in slightly different weights, having the same charge but different masses. These 'brothers' of nuclei are called *isotopes* (see also Note [6.15]).

[12.9]   We shall use the words 'fundamental' and 'elementary' to make a distinction that is not always made in the technical literature. On the subnuclear level all particles are fundamental. But particles that have no presently known further internal structure will be called elementary.

[12.10] The idea of the quark was first suggested in 1963 by Murray Gell-Mann and by George Zweig two theoretical physicists at the California Institute of Technology in Pasadena. The word is taken from the novel *Finnegans Wake* by James Joyce. The idea is taken from the symmetry group (see Note [3.8]) which is used to classify hadrons; it is the smallest mathematical entity involved there and Gell-Mann and Zweig suggested that this unit has more than formal mathematical importance.

[12.11] Top and bottom quarks are also called truth and beauty quarks.

[12.12] Among the best machines in current use is the Super Proton Synchrotron (SPS) at the European Center for Nuclear Research (CERN) in Geneva, Switzerland. That machine can accelerate protons to an energy that they would receive from a voltage jolt of 4×10^{11} volts. The Superconducting Super Collider (SSC) to be constructed in the United States is expected to produce two opposing proton beams where *each* proton would be accelerated corresponding to a voltage jolt of 2×10^{13} volts. And there are other very powerful particle accelerators under construction.

Glossary of technical terms

Some technical terms used in this book are *not* listed below. For these the dictionary definition is a good enough approximation for the present purpose.

acceleration: the time rate of change of velocity. It is a vector. If velocity is measured in m/s acceleration is measured in m/s^2.

alpha particle: a fundamental particle consisting of two protons and two neutrons. It can also serve as the atomic nucleus of the common form of the helium atom. These particles are emitted by certain heavy radioactive atomic nuclei.

amplitude: the deviation from the average in vibratory motion. Sometimes also used synonymously with 'maximum amplitude'.

angular momentum: the product of (linear) momentum times the distance of the line along which that momentum acts from the center of rotation. If a body has momentum p, its angular momentum is a measure of the rotation of that body about a given point, the center of rotation. Think of the plane through the center and through the line along which the vector p acts. Draw the distance d of that line from that center. Then the angular momentum L is pd. The axis of that rotation is the line through the center that is perpendicular to the plane. The direction of the angular momentum is by definition the sense along the axis in which a right-handed screw would progress if it were rotated the same way. The law of conservation of angular momentum preserves both its magnitude and its direction. The metric unit for L is Joule times second, i.e. it is 'energy' times 'time'.

classical physics: refers to those branches of physics that lie outside the field of quantum physics. It includes all the established theories known before the turn of this century as well as special and general relativity. Similarly: classical theory, classical particles, etc.

dynamics: a branch of physics dealing with the motion of bodies as a result of the action of given forces.

empiricism: a philosophical view asserting that all our knowledge (with the possible exception of logic and mathematics) derives from experience.

energy: a quantity that has many different forms of which the one easiest to conceptualize is work (see below). Other forms include kinetic, potential, heat, chemical, electric, gravitational, and mass energy. The law of conservation of energy permits any change from one form to another as long as the total energy is conserved. If a particular form of energy can be completely converted into work then this form can be thought of as capacity for doing work.

ensemble: a large collection of physical systems of the same constitution that may not all be in the same physical state such as a collection of atoms or of molecules with different velocities.

epistemology: a branch of philosophy dealing with the origin and validity of knowledge; the associated adjective is 'epistemological'.

epistemic: relating to knowledge.

event: a point in space and time. It refers to the location as well as to the instant of something taking place.

field: a region of space in which each point has a certain characteristic. It can refer to temperature or force or other characteristics. A temperature field is a region of space in which each point has a specific temperature. A force field is a region where there is a force (a vector) associated with each point.

force: a quantity characterizing push or pull. It is a vector having a magnitude as well as a direction.

frequency: number of vibrations per unit time. It is usually given per second in which case the unit is called 'hertz' after the German physicist Heinrich Hertz. 60 hertz means 60 vibrations per second.

holism: the belief that the whole cannot always be fully understood in terms of its parts, that there is more to a structured system than the sum of its parts.

group of transformations: a mathematical group (defined in Note [3.8]) whose elements are transformations.

index of refraction: when light propagates through a medium (water, glass, etc.) its speed, u, is less than in a vacuum where it is c. The ratio c/u is denoted by n and is called the index of refraction of that particular medium. It is determined by experiments.

inertial reference frame: a reference frame with respect to which the law of inertia is valid (that the absence of a force implies uniform motion).

instrumentalism: a philosophical view asserting that reality is assured only in the directly observed facts, and that theory plays the role of a tool, an instrument, that relates these facts. No commitment to the truth of theories is made, nor can one ascribe reality to its constructs.

interference: the superposition of two waves of equal wave lengths resulting in a wave of larger (constructive interference) or smaller (destructive interference) amplitude.

invariance of a law or set of laws: the property of remaining unchanged while certain quantities that describe the system *are* changed (transformed). Example: Newton's laws for the motion of a body subject to a force remain invariant when the positions of that body relative to one inertial reference frame are changed to the positions relative to another inertial frame (Galilean transformation).

light cone: the three-dimensional surface of a double cone in Minkowski space (Section 6f) formed by light emitted from one point into the future and past directions of time.

mass: a measure of a body's inertia, of its resistance to being accelerated. Its metric unit is the kilogram which is the mass of a cube of water whose sides are each 0.1 meters long.

momentum (short for 'linear momentum' as distinguished from 'angular momentum'): a vector quantity that is a measure of the thrust of a moving object. The rate at which momentum changes in time equals the force acting on the object. Momentum is conserved for a whole system if no forces act on the system from the outside. In Newtonian mechanics for a massive object whose mass does not change (not a jet plane that loses mass by ejecting exhaust gases), the momentum is just mass times velocity.

ontic: related to existence, to being.

ontology: a branch of philosophy dealing with existence and being; the associated adjective is 'ontological'.

phase: the relative position within one full period along the axis of a wave. Two waves that combine 'in phase' superpose with the same phase and will interfere constructively; 'out of phase' they superpose with phases differing by half a wavelength and interfere destructively.

photon: the smallest possible amount of monochromatic electromagnetic radiation. It has energy and momentum and behaves in every respect like a particle. But it has no mass. Being 'a piece of radiation' it does of course have a frequency and a wave length. Its velocity is that of the radiation.

probability amplitude: a quantity ψ whose square (see Note [10.6]) gives the probability. It can refer to probability of position, momentum, energy, or other attributes of a quantum particle.

quantum physics: the branch of physics that deals with quantum phenomena. It includes quantum mechanics, quantum optics, quantum electrodynamics, quantum field theory, and other areas.

realism: a philosophical view asserting the reality of abstract terms asserting that established scientific theories are at least approximately true and in this sense refer to real, existing entities (scientific realism).

reductionism: a philosophical view asserting the possibility of reducing a theory to a more fundamental theory in the sense that it can be deduced from the latter.

uniform motion: motion with a velocity that is constant in both magnitude and direction.

vector: a quantity such as force that has both magnitude and direction.

velocity: a vector whose magnitude is speed and whose direction is the direction of motion.

velocity of light: it plays a special role because its magnitude, the speed of light, is a universal constant in a vacuum. It is denoted by c and its exact value is by international convention 299 792 458 km/s. Given the international definition of the second, this indirectly provides for a new international definition of the meter which is more accurate than the older one (the meter stick in Paris).

wavelength: the distance to two neighboring crests of a wave.

worldline: the line in a space–time diagram showing the change of position in time of a pointlike object.

Name index

Subject index

aberration, 53
absolute space, 38ff
absolute time, 40f
acceleration, 217
acceleration of gravity, 212
acceptability of scientific theories, 17ff
accepted theories, 19
addition of velocities, 56, 71, 114
aim of scientific theories, 23–31
Almagest, 89
alpha particle, 217
amplitude, 217
antimatter, 192
approximation, 6–8, 11, 24, 51, 62, 113
Aristotelean logic, 181ff
atomic physics, 195
authority, 16, 17

baryons, 197
beauty in scientific theories, 13, 19
bending of light rays, 95–7
black-body radiation, 126–8, 154
black holes, 107–9
Bohr's atomic model, 132, 133, 213, 214
boost transformations, 46, 58, 59

caloric theory, 11
causality, 147
center of gravity, 208
chaotic systems, 145
classical physics, 121, 217
collapse of ψ, 173
color, 199
complementarity, 151–3
Compton effect, 131
confirmation of scientific theories, 17
conservation of energy, 72ff
conservation laws, 46, 47, 208
conservation of mass, 72
consistency, 18
contraction hypothesis, 57, 58, 62
conversion of mass into energy, 72ff
coordinate system, 104
Copenhagen interpretation, 25, 178, 180, 204
Copernican revolution, 16, 134

Copernican system, 40
correspondence principle, 153–6
curved space–time, 97–102

Darwinian evolution, 15, 134
determinism, 142, 144, 160, 179
diffraction, 215
double-slit experiment, 137–44, 155
dynamics, 217

Eddington's tables, 205
Einstein's general theory of relativity, see
    Einstein's gravitation theory
Einstein's gravitation theory, 20, 89–110, 121, 203
Einstein's principle of relativity, 59
Einstein's special theory of relativity, see special
    theory of relativity
Einstein–Podolsky–Rosen experiment, 172–7
electric field, 49
electrodynamics, 153
electromagnetic force, 29
electromagnetic spectrum, 29, 50
electromagnetic theory, 23, 28, 55
electromagnetic waves, 50, 128, 129
electromagnetism, 50
electron, 209
electron diffraction, 137
electron microscope, 205
electroweak theory, 30, 193
elementary particles, 196–201
empiricism, 207, 217
energy, 72ff, 217
entanglement, 171, 175, 179
entropy, 217
EPR experiment, see Einstein–Podolsky–Rosen
    experiment
equivalence principle, 93–7
established theories, 19, 111–4, 117
ether hypothesis, 52–5, 114
event, 65, 83, 220
explanation, 23ff

Fermat's Principle, 9–10, 220
field, 49, 217, 220